SpringerBriefs in Food, Health, and Nutrition

Springer Briefs in Food, Health, and Nutrition present concise summaries of cutting edge research and practical applications across a wide range of topics related to the field of food science.

Editor-in-Chief
Richard W. Hartel
University of Wisconsin—Madison, USA

Associate Editors
David Rodriguez-Lazaro, *ITACyL, Spain*
J. Peter Clark, *Consultant to the Process Industries, USA*
David Topping, *CSIRO, Australia*
John W. Finley, *Baton Rouge, Louisiana, USA*
Yrjö Roos, *University College Cork, Cork, Ireland*

T0214556

More information about this series at http://www.springer.com/series/10203

Antonio Gálvez • María José Grande Burgos
Rosario Lucas López • Rubén Pérez Pulido

Food Biopreservation

 Springer

Antonio Gálvez
Health Sciences Department,
  Microbiology Division,
  Faculty Experimental Sciences
University of Jaen
Jaen, Spain

Rosario Lucas López
Health Sciences Department,
  Microbiology Division,
  Faculty Experimental Sciences
University of Jaen
Jaen, Spain

María José Grande Burgos
Health Sciences Department,
  Microbiology Division,
  Faculty Experimental Sciences
University of Jaen
Jaen, Spain

Rubén Pérez Pulido
Health Sciences Department,
  Microbiology Division,
  Faculty Experimental Sciences
University of Jaen
Jaen, Spain

ISSN 2197-571X   ISSN 2197-5728 (electronic)
ISBN 978-1-4939-2028-0   ISBN 978-1-4939-2029-7 (eBook)
DOI 10.1007/978-1-4939-2029-7
Springer New York Heidelberg Dordrecht London

Library of Congress Control Number: 2014950425

Printed on acid-free paper

Springer is part of Springer Science+Business Media (www.springer.com)

Contents

Chapter 1
Introduction

Microbal foodborne diseases are a constant concern to human health, as shown by annual statistics published by official institutions (http://www.cdc.gov/foodborne-burden/; http://www.efsa.europa.eu/en/efsajournal/pub/3129.htm). The globalization of the food market and the large-scale distribution and processing of raw materials and food products create new ecological niches to which microorganisms from different regions of the world may adapt, raising new problems that the food industry must solve. This trend increases as the food chain tends to be more complex in several ways, including transportation distance, processing steps, distribution of raw materials, and shelf life extension of the finished products. The increase of the more susceptible populations (e.g., the young, elderly, and immunocompromised individuals), the migration of populations from rural to urban areas, the overexploitation of natural resources (such as soil and water) and the climatic changes, are also factors to be taken into consideration. The food industry also has to satisfy the newer consumer habits. In the past years, there has been a growing demand of consumers for foods that are fresh-tasting, lightly preserved, ready-to-eat, and (possibly) with health-promoting effects. Consumer organizations are also more and more concerned about the quality of foods and the ways in which they are produced.

Although some traditional methods for food production and processing are being abandoned, there is also a growing interest in traditional foods, and in the adaptation of local production processes to an industrial scale without substantial loss of the original value. Several developing countries already benefit from modern food industry; however, in many others, there is an overwhelming need to enhance the availability of raw materials and to promote food processing on an industrial scale in order to meet the nutritional requirements of the population and to provide a minimal framework of food safety. The need to avoid economic losses due to microbial spoilage of raw materials and food products, to decrease the incidence of food-borne illnesses, and to meet the food requirements of the growing world population strengthen the relevance of preservation methods in the food industry. In this context, the preservation of foods by natural, biological methods may be a satisfactory

© The Author(s) 2014

A. Gálvez et al., *Food Biopreservation*, SpringerBriefs in Food, Health, and Nutrition, DOI 10.1007/978-1-4939-2029-7_1

approach to solve many of the current food-related issues. All these factors have stimulated scientific research to exploit natural weapons, either alone or in combination with novel food processing technologies, in the development of biopreservation strategies compatible with the latest changes in human habits and lifestyle.

Chapter 2
Natural Antimicrobials for Food Biopreservation

Biopreservation or biocontrol refers to the use of natural or controlled microbiota, or its antibacterial products to extend the shelf life and enhance the safety of foods (Stiles 1996). Since lactic acid bacteria (LAB) occur naturally in many food systems and have a long history of safe use in fermented foods, thus classed as Generally Regarded As Safe (GRAS), they have a great potential for extended use in biopreservation. Antimicrobial substances from other natural sources, such as antimicrobial proteins or peptides from animal secretions, or bioactive molecules from plant or animal defense systems have also been exploited in different ways for food biopreservation.

2.1 Bacterial Antagonism as a Fundamental for Biopreservation

Microbes often live in complex ecosystems where they must interact with the biotic and abiotic components of the environment. Bacterial populations must compete for space and nutrients in order to survive. They have evolved different mechanisms such as nutrient and space competition, metabolic specialization, o cell differentiation, among others. One of the most common strategies to defend a population territory is ammensalism, which is based on the modification of the environment by the release of antimicrobial substances that inhibit growth or even kill competitors. Bacteria may release a variety of antimicrobial substances as byproducts of their normal metabolic activity. They also may produce more specific, dedicated antimicrobial weapons encoded by specific genetic determinants aimed specifically at combating other microbes. Metabolic products as well as antimicrobial peptides from lactic acid bacteria (LAB) have attracted great attention for food biopreservation. Being naturally or intentionally present in food fermentations, the lactic acid bacteria are considered themselves as natural preservatives as well as factories of

© The Author(s) 2014
A. Gálvez et al., *Food Biopreservation*, SpringerBriefs in Food, Health, and Nutrition, DOI 10.1007/978-1-4939-2029-7_2

natural antimicrobials for food biopreservation. LAB may produce a wide variety of active antagonistic metabolites such as organic acids (lactic, acetic, formic, propionic, butyric, hydroxyl-phenyllactic acid, and phenyllactic acid), diverse antagonistic compounds (carbon dioxide, ethanol, hydrogen peroxide, fatty acids, acetoin, diacetyl, reuterin, reutericyclin), antifungal compounds (propionate, phenyl-lactate, hydroxyphenyl-lactate, cyclic dipeptides, phenyllactic acid and 3-hydroxy fatty acids), and bacteriocins (such as nisin, pediocins, lacticins, enterocins and many others) (Muhialdin et al. 2011; Reis et al. 2012; Oliveira et al. 2014). Other bacterial groups (especially those from genus *Bacillus*) are also attracting attention because of the diversity of antimicrobial peptides they produce, some of which could also be exploited as biopreservatives.

2.1.1   Antimicrobial Substances Derived from Bacterial Cell Metabolism

Organic acids. Fermentation is an oxido-reductive process in which organic acids are the main end products. Lactic acid is the main organic acid produced during fermentation. Organic acids decrease the pH of the surrounding environment, creating a selective barrier against non-acidophiles. In addition, organic acids also have antibacterial activity. The antimicrobial effect of lactic acid is exerted by disruption of the cytoplasmic membrane and interference with membrane potential (Axe and Bailey 1995) and/or reduction in intracellular pH (Shelef 1994).

CO_2. Heterofermentative LAB produce CO_2 as a byproduct of sugar fermentation. The production of CO_2 creates an anaerobic environment has antagonistic effects specifically against aerobic bacteria (Adams and Nicolaides 1997). It also dissolves in water, generating carbonic acid.

Diacetyl. Certain LAB may produce diacetyl (2,3-butanedione) is a by-product of the metabolic activity (Jay 1982). Diacetyl exhibits antibacterial activity against *Listeria, Salmonella, Escherichia coli, Yersinia,* and *Aeromonas* (Jay 1982). Gram-negative bacteria are generally more sensitive than gram-positive bacteria to diacetyl (Adams and Nicolaides 1997). However, since the high concentrations of diacetyl required to achieve inhibition of spoilage bacteria also affect the sensory properties of the food (Helander et al. 1997), the use of diacetyl-producing cultures for protective purposes should be limited to foods where diacetyl is an essential component of the food sensory properties (Jay 1982).

Hydrogen peroxide. LAB are deficient in catalase activity. Hydrogen peroxide is produced in the presence of oxygen as a result of the action of flavoprotein oxidases or NADH peroxidase (Ammor et al. 2006). Hydrogen peroxide is thought to elicit an antibacterial effect through oxidative damage of proteins, but it also may increase membrane permeability (Kong and Davison 1980).

Reuterin. Also known as 3-hydroxypropionaldehyde (3-HPA), reuterin is a low-molecular-weight antimicrobial compound produced by *Lactobacillus reuteri*

(Talarico et al. 1988) and some other LAB. It is formed as an intermediate during the metabolism of glycerol to 1,3-propanediol under anaerobic conditions (Talarico et al. 1988; Talarico and Dobrogosz 1989) and behaves in solution as an equilibrium mixture of monomeric, hydrated monomeric and cyclic dimeric forms of 3-HPA. The antimicrobial activity of reuterin has been attributed to its ability to inhibit DNA synthesis (Talarico and Dobrogosz 1989), being active on bacteria as well as yeasts and molds. Reuterin has broad spectrum of activity and inhibits fungi, protozoa and a wide range of bacteria including both gram-positive and gram-negative bacteria. It has bacteriostatic activity against *L. monocytogenes* and variable bactericidal activities against *Staphylococcus aureus*, *E. coli* O157:H7, *Salmonella* Choleraesuis, *Yersinia enterocolitica*, *Aeromonas hydrophila* subsp. *hydrophila,* and *Campylobacter jejuni* (Arqués et al. 2004).

Reutericyclin. This unique tetramic acid is a negatively charged, highly hydrophobic antagonist (Gänzle et al. 2000). Reutericyclin acts as a proton ionophore, resulting in dissipation of the proton motive force (Gänzle 2004). It lacks activity towards yeasts and fungi, but it is active on gram-positive bacteria including *Lactobacillus* spp., *Bacillus subtilis, Bacillus cereus, Enterococcus faecalis, S. aureus* and *Listeria innocua.* Spore germination of *Bacillus* species was inhibited by this antimicrobial compound, but the spores remained unaffected under conditions that do not permit germination. As in many other antagonists, inhibition of Gram-negative bacteria (*E. coli* and *Salmonella*) is observed under conditions that disrupt the outer membrane, including truncated lipopolysaccharides (LPS), low pH and high salt concentrations. Reutericyclin was shown to be produced in concentrations active against competitors during growth of *Lactobacillus reuteri* in sourdough. It was proposed that reutericyclin-producing strains may have applications in the biopreservation of foods (Gänzle 2004).

2.1.2   Antifungal Compounds

The only antifungal compound approved for food applications is derived from an actinomycete. Natamycin (pimaricin) is an antifungal compound produced by *Streptomyces natalensis,* approved as a broad-spectrum antifungal biopreservative for foods and beverages (Stark 2003). Natamycin binds irreversibly to the cell membrane of fungi because of its high affinity for ergosterol. This causes membrane hyperpermeability leading to rapid leakage of essential ions and peptides and ultimately cell lysis (Teerlink et al. 1980). There is a growing interest in antifungal compounds from LAB, and a few LAB strains showing antifungal activities have been characterized regarding the antimicrobial substances responsible for the observed inhibitions. Usually, antifungal activity produced by LAB is due to the combination of organic acids and cyclic dipeptides. Besides lactic acid, other organic molecules such as phenyllactic and 4-hydroxy-phenyllactic acids (Dal Bello et al. 2007; Lavermicocca et al. (2000, 2003)), benzoic acid, methylhydantoin, mevalonolactone (Niku-Paavola et al. 1999), 5-oxododecanoic acid, 3-hydroxy decanoic acid and 3-hydroxy-5-dodecenoic acid (Ryu et al. 2014) or 3,6-bis(2-methylpropyl)-2,5-piperazinedion (Yang and Chang 2010) have been identified. Among the cyclic dipeptides described in cell-free supernatants of antifungal LAB strains are cyclo (Gly-LLeu) (Niku-Paavola et al. 1999), cyclo (L-Leu-L-Pro) and cyclo (L-Phe-L-Pro) (Dal Bello et al. 2007), cyclo (L-Phe-l-Pro) and cyclo (L-Phe-trans-4-OH-L-Pro) (Ström et al. 2002). Some of them require high concentrations in order to be effective, while others have demonstrated potential for biopreservation in certain foods such as bread (Dal Bello et al. 2007).

2.1.3 Bacteriocins

Bacteriocins can be defined as ribosomally synthesized antimicrobial peptides or proteins, which can be posttranslationally modified or not (Jack et al 1995). Bacteriocins from Gram-positive bacteria are generally classified according to size, structure, and modifications. Klaenhammer (1993) defined four classes of bacteriocins produced by LAB. Class I bacteriocins or 'lantibiotics' are small, ribosomally synthesized peptides that undergo extensive post-translational modification. They contain lanthionine and b-methyl lanthionine residues, as well as dehydrated amino acids. Class II bacteriocins are small (4–6 kDa), heat-stable, ribosomally synthesized peptides which were differentiated from lantibiotics because they do not undergo extensive post-translational modification, except for cleavage of a leader peptide (when present) during transport out of the cell. Nevertheless, some exceptions to this rule have been reported recently, illustrated by the n-terminal formylated two-peptide bacteriocin from *E. faecalis* 710C (Liu et al. 2011) and enterocin BacFL31 which contains hydroxyproline residues (Chakchouk-Mtibaa et al. 2014). Nes et al. (1996) regrouped the class II bacteriocins, retaining class IIa and IIb but changing class IIc to include bacteriocins that contain a typical signal peptide and that are secreted by the general translocase (*sec*) pathway of the cell. Cotter et al. (2005) suggested to divide class II bacteriocins into several subclasses: class IIa (pediocin-like bacteriocins), class IIb (two-peptide bacteriocins), and class IIc (circular bacteriocins) Cotter et al. (2005). However, circular bacteriocins may also be considered as a separate class (Franz et al. 2007; van Belkum et al. 2011). Nonbacteriocin lytic proteins, termed bacteriolysins (also referred to as class III bacteriocins), are large and heat-labile proteins with a distinct mechanism of action from other Gram-positive bacteriocins (Cotter et al. 2005). Specific classification schemes were also proposed for bacteriocins from genus *Enterococcus* (Franz et al. 2007) and genus *Bacillus* (Abriouel et al. 2011). This last one generates a wide variety of peptide structures containing modified amino acid residues other than the classical ones found in, for example, nisin. An updated classification of bacteriocins was proposed by Rea et al. (2011) including two additional subclasses for the lantibiotics and one additional subclass for the non-modified peptides. An orientative summary on the diversity of bacteriocins from Gram-positive bacteria is presented in Table 2.1.

Table 2.1 Classes of bacteriocins produced by Gram-positive bacteria

Class	Features	Type or subclass	Examples
Class I (lantibiotics)	Posttranslational modification yielding unusual amino acid residues	Type A: Cationic, amphiphilic, pore forming activity on the bacterial membrane	Nisin A/Z, lacticin 481, lacticin 3147, subtilin, plantaricin C, varicin 8, lactocin S
		Type B: Globular, no or negative charge, inhibit phospholipase A2	Mersacidin, duramycin B/C, cinnamycin
		Labyrinthopeptides: contain labionin, a carbocyclic, post-translationally modified amino acid residue	Labyrinthopeptins A1, A2, and derivatives
		Sactibiotics: Contain intramolecular sulfur to a-carbon crosslinkages	Subtilosin A, thuricin CD
Class II (nonlantibiotic peptides)	No posttranslational modification, small cationic, amphiphilic peptides[a]	Class IIa: Antilisterial, pediocin-like bacteriocins (YGNGV motif)	Pediocin AcH/PA-1, sakacin A, curvacin A, enterocin A
		Class IIb: Two-peptide bacteriocins[b]	Enterocin L50, lacticin F, lactococcin G and Q; plantaricins EF and JK
		Class IIc: Circular bacteriocins	Enterocin AS-48, reutericin 6, gassericin A, lactocyclin Q
		Class IId: Other single-peptide, nonpediocin molecules[b]	Lactococcins A and 972, enterocin EJ97, divergicin A
Class III (bacteriolysins)	Nonbacteriocin lytic proteins, large, heat labile, cause cell lysis through cell wall hydrolysis		Lysostaphin, helveticin J, enterolysin A

[a]Some may contain modified amino acid residues like n-terminal formylation or hydroxyproline
[b]Some of these are synthesized without leader peptide, and could be included in a separate class of leaderless bacteriocins

2.2 Antimicrobials from Animal Sources

Antimicrobial proteins and peptides are naturally found as part of the defense system of living organisms (including humans, animals, plants, insects…). Lysozyme, lactoferrin and ovotransferrin are illustrative examples. Lysozyme from different sources is commercialized as a natural preservative for food applications, either singly or in combination with other antimicrobials. Lysozyme is generally recognized as safe (GRAS) for direct addition to foods (FDA 1998). Lactoferrin (and its partial hydrolysis derivative lactoferricin) is another natural protein (which is found in milk and other secretions) with antimicrobial activity due to its iron-binding capacity and polycationic nature (Ellison 1994). Lactoferrin shows antimicrobial activity against a wide range of bacteria (including foodbone pathogens like *Carnobacterium, L. monocytogenes, E. coli*, and *Klebsiella*) and viruses (Lönnerdal 2011; Gyawali and Ibrahim 2014), and has been approved for application on beef in the United States and has been applied as an antimicrobial in a variety of meat products (Juneja et al. 2012; USDA-FSIS 2010).

Lactoperoxidase is another antimicrobial system that originated from milk and is reported to be effective against gram-negative bacteria (de Wir and van Hooydonk 1996). Ovotransferrin has a high affinity for iron, and inhibits bacterial growth due to iron deprivation (Valenti et al. 1987). Interestingly, hydrolysis of natural proteins may yield peptide fragments with diverse biological activities, including antimicrobial activity (Möller et al. 2008). Following a strategy of "tailoring and modelling," a number of short peptides with high bactericidal activity have been developed from the bactericidal domain of lysozyme. Ovotransferrin, alpha-lactalbumin and beta-lactoglobulin have also been investigated as sources of antimicrobial peptides (Pellegrini 2003).

Protamine is composed of cationic antimicrobial peptides naturally present in spermatic cells of fish, birds and mammals (Rodman et al. 1984) and is commercially recovered from herring (clupeine) and salmon (salmine) milt. With a MW of 4,112 Da and a pI of 11–13, protamine is the most cationic naturally occurring cationic antimicrobial peptide described to date (Potter et al. 2005). It shows broad antimicrobial activity against gram-positive bacteria, gram-negative bacteria, and fungi (Uyttendaele and Debevere 1994). Protamine has been used to preserve a wide variety of foods ranging from confection items to fruits and rice.

Pleurocidin is present in myeloid cells and mucosal tissues of many vertebrates and Invertebrates (Jia et al. 2000). It shows antimicrobial activity against several foodborne bacteria, such as *L. monocytogenes* and *E. coli* O157:H7, and pathogenic fungi (Burrowes et al. 2004; Jung et al. 2007).

Chitosan is a polycationic biopolymer naturally present in the exoskeletons of crustaceans and arthropods (Tikhonov et al. 2006). Partially and fully deacetylated chitosan derivatives of low molecular weight are available, with broad antibacterial and antifungal activity (Franklin and Snow 1981; Kong et al. 2010). Chitosan is considered a safe food additive. Reported antibacterial activity for chitosan derivatives include both Gram-positive and Gram-negative bacteria such as *S. aureus, L. monocytogenes, B. cereus, E. coli, Shigella dysenteriae*, and *Salmonella* Typhimurium (Gyawali and Ibrahim 2014). Chitosan has attracted great attention for development of biodegradable edible coatings, singly or dosed with other antimicrobial substances (Maher et al. 2013).

2.3 Antimicrobials Derived from Plants

Herbs and spices have been recognized to possess a broad spectrum of active constituents that exhibit antibacterial, antifungal, antiparasitic, and/or antiviral activities. Essential oils have been used for centuries as part of natural traditional medicine. They are aromatic oily liquids obtained from plant material (flowers, buds, seeds, leaves, twigs, bark, herbs, wood, fruits and roots). The major groups of principal components that make essential oils effective antimicrobials include saponins, flavonoids, carvacrol, thymol, citral, eugenol, linalool, terpenes, and their precursors (Burt 2004). The antimicrobial activity of alliums is mainly attributed to various kinds of alk(en)yl alka/ene thiosulfinates (thiosulfinates; and their transformation products (Kyung 2012). Allium-derived antimicrobial compounds inhibit microorganisms by reacting with the sulfhydryl (SH) groups of cellular proteins. In olive oil, distinctive antimicrobial compounds including oleuropein, oleuropein aglycon, elenoic acid and oleocanthal (in addition to hydroxytyrosol and tyrosol) have been described (Cicerale et al. 2012). The antibacterial activities of essential oils and other plant extracts has attracted great attention for application of the crude extracts or their bioactive components in food biopreservation (Burt 2004; Holley and Patel 2005; Richard and Patel 2005; Tajkarimi et al. 2010). In the concentration range of 0.05–0.1 %, essential oils have demonstrated activity against pathogens, such as *S*. Typhimurium, *E. coli* O157:H7, *L. monocytogenes*, *B. cereus* and *S. aureus*, in food systems. However, activity against various microorganisms on food products might be higher than the concentration applied for flavoring purposes. As a result, this might cause food tainting and/or adverse sensorial effects to food products (Bagamboula et al. 2004). It has been suggested that the adverse sensorial effects of essential oils agents to food products can be overcome by masking the odor with other approved aroma compounds (Gutiérrez et al. 2009). Antimicrobial agents derived from essential oils are interesting candidates for development of activated films or packagings.

Plants also produce a variety of antimicrobial peptides, many of which can be grouped in different classes: thionins, defensins, lipid transfer proteins, cyclotides and snakins (Padovan et al. 2010). Some of them could possibly be exploited for food biopreservation. Interestingly, plants can be a good source of antifungal proteins and peptides, including chitinases, glucanases, thaumatin-like proteins, thionins, and cyclophilin-like proteins (Ng 2004).

2.4 Bacteriophages

Bacteriophages are obligate parasites of bacteria. Lytic bacteriophages offer a great potential as natural biopreservative agents, due to their capacity to selectively control bacterial populations. This phenomenon occurs spontaneously in Nature, but can also be applied purposedly in food systems. One feature of bacteriophages is their high host specificity, at the level of species and even strains. Specificity at strain level can be a limitation for application of bacteriophages. Nevertheless, several studies have shown the efficacy in food systems of mixtures containing different bacteriophages and also broad-host range bacteriophages that were able to attack a high number of bacterial strains, including the most virulent strains found in foods (Hagens and Loessner 2007, 2010, 2014; Sharma 2013; Sulakvelidze 2013). Control of *L. monocytogenes* by bacteriophages has been addressed in many different ready-to-eat foods of animal as well as plant origin. Commercial phage preparations like ListShield™ (containing a mixture of six naturally occurring listeriophages) and Listex™ P100 (based on listeriophage P100) have been approved by the FDA and USDA (Hagens and Loessner 2014). Bacteriophages specific for *Salmonella* serotypes have also been used on various food substrates such as sprouted seeds and animal skins and carcasses. The commercial preparation SalmoFresh contains a cocktail of naturally occurring lytic bacteriophages that selectively and specifically kill *Salmonella*, including strains belonging to the most common/highly pathogenic serotypes: Typhimurium, Enteritidis, Heidelberg, Newport, Hadar, Kentucky, and Thompson. Bacteriophages specific for *E. coli* (including virulent strains) have also shown efficacy on different food substrates. The commercial preparation EcoShield™ contains a cocktail of three lytic phages specific for *E. coli* O157:H7. Illustrative studies on application of bacteriophages can be found in the scientific literature for other human pathogenic or toxinogenic bacteria such as *Shigella* spp. (Zhang et al. 2013), *C. jejuni* (Bigwood et al. 2008), *Cronobacter sakazakii* (Zuber et al. 2008), *S. aureus* (Bueno et al. 2012), as well as spoilage bacteria such as *Pseudomonas fluorescens* (Sillankorva et al. 2008). *Brochothrix thermosphacta* (Greer and Dilts 2002) or *Leuconostoc gelidum* (Greer et al. 2007).

Another emerging field of interest is the application of bacteriophages for reducing the carriage of zoonotic agents in livestock and poultry and also for the prophilaxy and therapy in diseased animals. Phage therapy is potentially useful in virulent *Salmonella* and *E. coli* infections in chickens, calves and pigs, and in control of the food-borne pathogens *Salmonella* and *C. jejuni* in chickens and *E. coli* O157:H7 in cattle (Johnson et al. 2008; Connerton et al. 2011; Sulakvelidze 2013; Endersen et al. 2014). Selective application of bacteriophages could improve animal health and animal production and reduce the risks of transmission of zoonotic agents to humans.

References

Abriouel H, Franz CMAP, Ben Omar N (2011) Diversity and Applications of *Bacillus* Bacteriocins. FEMS Microbiol Revs 35:201–232

Adams MR, Nicolaides L (1997) Review of the sensitivity of different foodborne pathogens to fermentation. Food Control 8:227–239

Ammor S, Tauveron G, Dufour E et al (2006) Antibacterial activity of lactic acid bacteria against spoilage and pathogenic bacteria isolated from the same meat small-scale facility. 1. Screening and characterization of the antibacterial compounds. Food Control 17:454–461

Arqués JL, Fernandez J, Gaya P et al (2004) Antimicrobial activity of reuterin in combination with nisin against food-borne pathogens. Int J Food Microbiol 95:225–229

Axe DD, Bailey JE (1995) Transport of lactate and acetate through the energized cytoplasmic membrane of *Escherichia coli*. Biotechnol Bioeng 47:8–19

Bagamboula CF, Uyttendaele M, Debevere J (2004) Inhibitory effect of thyme and basil essential oils, carvacrol, thymol, estragol, linalool and *p*-cymene towards *Shigella sonnei* and *S. flexneri*. Food Microbiol 21:33–42

Bigwood T, Hudson JA, Billington C et al (2008) Phage inactivation of foodborne pathogens on cooked and raw meat. Food Microbiol 25:400–406

Bueno E, García P, Martínez B et al (2012) Phage inactivation of *Staphylococcus aureus* in fresh and hard-type cheeses. Int J Food Microbiol 158:23–27

Burrowes O, Hadjicharalambous C, Diamond G et al (2004) Evaluation of antimicrobial spectrum and cytotoxic activity of pleurocidin for food applications. J Food Sci 69:FMS66–FMS71

Burt S (2004) Essential oils: their antibacterial properties and potential applications in foods – a review. Int J Food Microbiol 94:223–253

Chakchouk-Mtibaa A, Elleuch L, Smaoui S et al (2014) An antilisterial bacteriocin BacFL31 produced by *Enterococcus faecium* FL31 with a novel structure containing hydroxyproline residues. Anaerobe 27:1–6

Cicerale S, Lucas LJ, Keast RSJ (2012) Antimicrobial, antioxidant and anti-inflammatory phenolic activities in extra virgin olive oil. Curr Opin Biotechnol 23:129–135

Connerton PL, Timms AR, Connerton IF (2011) *Campylobacter* bacteriophages and bacteriophage therapy. J Appl Microbiol 111:255–265

Cotter PD, Hill C, Ross RP (2005) Bacteriocins: developing innate immunity for food. Nat Rev Microbiol 3:777–7788

Dal Bello F, Clarke C, Ryan L et al (2007) Improvement of the quality and shelf life of wheat bread by fermentation with the antifungal strain *Lactobacillus plantarum* FST 1.7. J Cereal Sci 45:309–318

de Wir J, van Hooydonk A (1996) Structure, functions and applications of lactoperoxidase in natural antimicrobial systems. Neth Milk Dairy 50:227–244

Ellison RT (1994) The effects of lactoferrin on Gram-negative bacteria. In: Hutchens TW, Lönnerdal B, Rumball S (eds) Lactoferrin-structure and function. Plenum Press, New York, pp 71–87

Endersen L, O'Mahony J, Hill C et al (2014) Phage therapy in the food industry. Annu Rev Food Sci Technol 5:327–349

FDA (1998) Direct food substances affirmed as generally recognized as safe: egg white lysozyme. Fed Register 63:12421–12426

Franklin TJ, Snow GA (1981) Biochemistry of antimicrobial action, 3rd edn. Chapman and Hall, London

Franz CMAP, van Belkum MJ, Holzapfel WH (2007) Diversity of enterococcal bacteriocins and their grouping into a new classification scheme. FEMS Microbiol Rev 31:293–310

Gänzle MG (2004) Reutericyclin: biological activity, mode of action, and potential applications. Appl Microbiol Biotechnol 64:326–332

Gänzle MG, Höltzel A, Walter J et al (2000) Characterization of reutericyclin produced by *Lactobacillus reuteri* LTH2584. Appl Environ Microbiol 66:4325–4333

Greer GG, Dilts BD (2002) Control of *Brochothrix thermosphacta* spoilage of pork adipose tissue using bacteriophages. J Food Prot 65:861–863

Greer GG, Dilts BD, Ackermann HW (2007) Characterization of a *Leuconostoc gelidum* bacteriophage from pork. Int J Food Microbiol 114:370–375

Gutiérrez L, Escudero A, Rn B et al (2009) Effect of mixed antimicrobial agents and flavors in active packaging films. J Agric Food Chem 57:8564–8571

Gyawali R, Ibrahim SA (2014) Natural products as antimicrobial agents. Food Control 46:412–429

Hagens S, Loessner MJ (2007) Application of bacteriophages for detection and control of foodborne pathogens. Appl Microbiol Biotechnol 76:513–519

Hagens S, Loessner MJ (2010) Bacteriophage for biocontrol of foodborne pathogens: calculations and considerations. Curr Pharm Biotechnol 11:58–68

Hagens S, Loessner MJ (2014) Phages of *Listeria* offer novel tools for diagnostics and biocontrol. Front Microbiol 10:5–159

Helander IM, von Wright A, Mattila-Sandholm TM (1997) Potential of lactic acid bacteria and novel antimicrobials against Gram-negative bacteria. Trends Food Sci Technol 8:146–150

Holley RA, Patel D (2005) Improvement in shelf-life and safety of perishable foods by plant essential oils and smoke antimicrobials. Food Microbiol 22:273–292

Jack RW, Tagg JR, Ray B (1995) Bacteriocins of Gram positive bacteria. Microbiol Rev 59:171–200

Jay JM (1982) Antimicrobial properties of diacetyl. Appl Environ Microbiol 44:525–532

Jia X, Patrzykat A, Devlin RH et al (2000) Antimicrobial peptides protect Coho salmon from *Vibrio anguillarum* infections. Appl Environ Microbiol 66:1928–1932

Johnson RP, Gyles CL, Huff WE et al (2008) Bacteriophages for prophylaxis and therapy in cattle, poultry and pigs. Anim Health Res Rev 9:201–215

Juneja VK, Dwivedi HP, Yan X (2012) Novel natural food antimicrobials. Annu Rev Food Sci Technol 3:381–403

Jung HJ, Park Y, Sung WS et al (2007) Fungicidal effect of pleurocidin by membrane-active mechanism and design of enantiomeric analogue for proteolytic resistance. Biochim Biophysic Acta (BBA)-Biomembr 1768:1400–1405

Klaenhammer TR (1993) Genetics of bacteriocins produced by lactic acid bacteria. FEMS Microbiol Rev 12:39–86

Kong S, Davison AJ (1980) The role of interactions between O_2, H_2O_2, $\cdot OH$, e^- and O_2^- in free radical damage to biological systems. Arch Biochem Biophys 204:18–29

Kong M, Chen XG, Xing K, Park HJ (2010) Antimicrobial properties of chitosan and mode of action: a state of the art review. Int J Food Microbiol 144:51–63

Kyung KH (2012) Antimicrobial properties of allium species. Curr Opin Biotechnol 23:142–147

Lavermicocca P, Valerio F, Evidente A et al (2000) Purification and characterization of novel antifungal compounds from the sourdough *Lactobacillus plantarum* strain 21B. Appl Environ Microbiol 66:4048–4090

Lavermicocca P, Valerio F, Visconti A (2003) Antifungal activity of phenyllacti acid against molds isolated from bakery products. Appl Environ Microbiol 69:634–640

Liu X, Vederas JC, Whittal RM et al (2011) Identification of an N-terminal formylated, two-peptide bacteriocin from *Enterococcus faecalis* 710C. J Agric Food Chem 59:5602–5608

Lönnerdal B (2011) Biological effects of novel bovine milk fractions. Nestle Nutr Workshop Ser Paediatr Program 67:41–54

Maher Z, Entsar E, Abdou S (2013) Chitosan based edible films and coatings: a review. Mat Sci Eng C 33:1819–1841

Möller NP, Scholz-Ahrens KE, Roos N et al (2008) Bioactive peptides and proteins from foods: indication for health effects. Eur J Nutr 47:171–182

Muhialdin BJ, Hassan Z, Sadon SK (2011) Biopreservation of food by lactic acid bacteria against spoilage fungi. Ann Food Sci Technol 12:45–57

Nes IF, Diep DB, Håvarstein LS et al (1996) Biosynthesis of bacteriocins in lactic acid bacteria. Antonie Van Leeuwen 70:113–128

Ng TB (2004) Antifungal proteins and peptides of leguminous and non-leguminous origins. Peptides 25:1215–1222

Niku-Paavola ML, Laitila A, Mattila-Sandholm T et al (1999) Newtypes of antimicrobial compounds produced by *Lactobacillus plantarum*. J Appl Microbiol 86:29–35

Oliveira PML, Zannini E, Arendt EK (2014) Cereal fungal infection, mycotoxins, and lactic acid bacteria mediated bioprotection: from crop farming to cereal products. Food Microbiol 37:78–95

Padovan L, Scocchi M, Tossi A (2010) Structural aspects of plant antimicrobial peptides. Curr Prot Pept Sci 11:210–219

Pellegrini A (2003) Antimicrobial peptides from food proteins. Curr Pharm Des 9:1225–1238

Potter R, Truelstrup Hansen L, Gill TA (2005) Inhibition of foodborne bacteria by native and modified protamine: importance of electrostatic interactions. Int J Food Microbiol 103:23–34

Rea MC, Ross P, Cotter PD et al (2011) Classifications of bacteriocins from Gram positive bacteria. In: Drider D, Rebuffat S (eds) Prokaryotic antimicrobial peptides: from genes to applications. Springer, New York, pp 29–53

Reis JA, Paula AT, Casarotti SN et al (2012) Lactic acid bacteria antimicrobial compounds: characteristics and applications. Food Eng Rev 4:124–140

Richard AH, Patel D (2005) Improvement in shelf-life and safety of perishable foods by plant essential oils and smoke antimicrobials. Food Microbiol 22:273–292

Rodman TC, Pruslin FH, Allfrey VG (1984) Protamine-DNA association in mammalian spermatozoa. Experim Cell Res 150:269–281

Ryu EH, Yang EJ, Woo ER et al (2014) Purification and characterization of antifungal compounds from *Lactobacillus plantarum* HD1 isolated from kimchi. Food Microbiol 41:19–26

Sharma M (2013) Lytic bacteriophages. Potential interventions against enteric bacterial pathogens on produce. Bacteriophage 3:e25518

Shelef LA (1994) Antimicrobial effects of lactates, a review. J Food Prot 57:445–450

Sillankorva S, Neubauer P, Azeredo J (2008) *Pseudomonas fluorescens* biofilms subjected to phage phiIBB-PF7A. BMC Biotechnol 27:8–79

Stark J (2003) Natamycin: an effective fungicide for food and beverages (p 82–95). In: Roller S (ed) Natural antimicrobials for the minimal processing of foods. Woodhead, Cambridge, pp 82–97

Stiles M (1996) Biopreservation by lactic acid bacteria. Antonie Van Leeuwen 70:331–345

Ström K, Sjögren J, Broberg A et al (2002) *Lactobacillus plantarum* MiLAB 393 produces the antifungal cyclic dipeptides cyclo (L-Phe-L-Pro) and cyclo (L-Phe-trans-4-OH-L-Pro) and 3-phenyllactic acid. Appl Environ Microbiol 68:4322–4327

Sulakvelidze A (2013) Using lytic bacteriophages to eliminate or significantly reduce contamination of food by foodborne bacterial pathogens. J Sci Food Agric 93:3137–3146

Tajkarimi MM, Ibrahim SA, Cliver DO (2010) Antimicrobial herb and spice compounds in food. Food Control 21:1199–1218

Talarico TL, Dobrogosz WJ (1989) Chemical characterization of an antimicrobial substance produced by *Lactobacillus reuteri*. Antimicrob Agents Chemother 33:674–679

Talarico TL, Casas IA, Chung TC et al (1988) Production and isolation of reuterin, a growth inhibitor produced by *Lactobacillus reuteri*. Antimicrob Agents Chemother 32:1854–1858

Teerlink T, de Kruijff B, Demel RA (1980) The action of pimaricin, etruscomycin and amphotericin B on liposomes with varying sterol content. Biochim Biophys Acta 599:484–492

Tikhonov VE, Stepnova EA, Babak VG et al (2006) Bactericidal and antifungal activities of a low molecular weight chitosan and its N-/2(3)-(dodec-2-enyl) succinoyl/-derivatives. Carbohyd Polym 64:66–72

USDA-FSIS (2010) Safe and suitable ingredients used in the production of meat, poultry, and egg products. FSIS Dir. 7120.1 Revision 2

Uyttendaele M, Debevere J (1994) Evaluation of the antimicrobial activity of protamine. Food Microbiol 11:417–427

Valenti P, Visca P, Antonini G et al (1987) The effect of saturation with Zn2+ and other metal ions on the antibacterial activity of ovotransferrin. Med Microbiol Immunol 176:123–130

van Belkum MJ, Martin-Visscher LA, Vederas JC (2011) Structure and genetics of circular bacte-
    riocins. Trends Microbiol 19:411–418
Yang EJ, Chang HC (2010) Purification of a new antifungal compound produced by *Lactobacillus
    plantarum* AF1 isolated from kimchi. Int J Food Microbiol 139:56–63
Zhang H, Wang R, Bao H (2013) Phage inactivation of foodborne *Shigella* on ready-to-eat spiced
    chicken. Poultry Sci 92:211–217
Zuber S, Boissin-Delaporte C, Michot L et al (2008) Decreasing *Enterobacter sakazakii*
    (*Cronobacter* spp.) food contamination level with bacteriophages: prospects and problems.
    Microb Biotechnol 1:532–543

Chapter 3
Application of Lactic Acid Bacteria and Their Bacteriocins for Food Biopreservation

3.1 Bacteriocins and Bactericin-Producing Strains

Microbes and/or their natural products have played key roles in the preservation of foods during mankind history (Ross et al. 2002). The rational exploitation of microbial antagonism based on scientific knowledge has been possible after the discovery of the biochemical nature of the antimicrobial substances produced by microorganisms. Bacteriocins produced by the lactic acid bacteria (LAB) have several features that still make them attractive for food preservation: (1) LAB have a long history of safe use in foods; (2) LAB and their cell products—including bacteriocins—are generally recognised as safe; (3) LAB bacteriocins are not active and non-toxic on eukaryotic cells, and (4) due to their proteinaceous nature, bacteriocins are expected to become inactivated by digestive proteases and not exert significant effects on gut microbiota at the concentrations ingested with the food. In addition, LAB bacteriocins may be suitable as preservatives, given (1) their sometimes broad antimicrobial spectrum, including food poisoning and spoilage bacteria, (2) their synergistic activity with other antimicrobials, (3) a bactericidal mode of action exerted at membrane level, which avoids cross resistance with antibiotics of clinical use, (4) stability under the heat and pH conditions achieved under processing of many foods, and (5) their genetic determinants are usually plasmid-encoded, which facilitates genetic manipulation and development of producer strains with improved technological properties. Bacteriocin-encoding plasmids may be transferred to other strains by natural processes, but at the same time there is a risk for loss of the plasmid together with the bacteriocin production capacity.

Application of bacteriocins in food preservation may be beneficial in several aspects (Thomas et al. 2000; Gálvez et al. 2007): (1), to decrease the risks of food poisoning; (2) to decrease cross contamination in the food chain; (3) improve the

© The Author(s) 2014
A. Gálvez et al., *Food Biopreservation*, SpringerBriefs in Food, Health, and Nutrition, DOI 10.1007/978-1-4939-2029-7_3

shelf life of food products; (4) food protection during temperature abuse episodes; (5) decrease economic losses due to food spoilage; (6) reduce the levels of added chemical preservatives; (7) reduce the intensity of physical treatments, achieving a better preservation of the food nutritional value and possibly decrease of processing costs; (8) may provide alternative preservation barriers for "novel" foods (less acidic, with a lower salt content, and with a higher water content), and (9) may satisfy the demands of consumers for foods that are fresh-tasting, lightly-preserved, and ready to eat (RTE). There may also be a potential market for bacteriocins as natural substitutes for chemical preservatives, and in the preservation of functional foods and nutraceuticals (Robertson et al. 2004).

According to previous studies (Deegan et al. 2006; Gálvez et al. 2007), bacteriocins can be applied in foods in many different ways:

(a) Addition of bacteriocin preparations. These are often obtained by cultivation of the producer strains in growth media suitable for bacteriocin production. The bacteriocin production step is usually followed by inactivation of the producer bacterial cells (for example by heat or UV treatments) and concentration of the cultured broths (by lyophilisation or spray-drying) to obtain a bio-active powder which contains a mixture of the antimicrobial substances produced in broth (such as the bacteriocin and organic acids). Commercial preparations such as Nisaplin™, Alta™ products or Microgard™ are some examples. Other bacteriocins such as lacticin 3147, variacin from *Kokuria varians* or enterocin AS-48 have also been obtained as dry powder preparations (Morgan et al. 1999; O'Mahony et al. 2001; Ananou et al. 2010b). However, most bacteriocins have only been produced on laboratory culture media, and recovered as partially purified concentrates by standard protein purification techniques such as ammonium sulphate precipitation or ion exchange chromatography.

(b) Bacteriocin-producing cultures. These may come in the form of lyophilized preparations for propagation and inoculum build up or direct addition to the food, as well as overnight cultures or even refrigerated cell concentrates. Bacteriocin-producing cultures must contain metabolically active bacterial cells ready for propagation in the food substrate and able to carry a rapid in situ bacteriocin production. Bacteriocin-producing strains can be applied as the main starter cultures in fermented foods provided that they offer the technological properties required for the fermentation, or as an adjunct culture in combination with bacteriocin-resistant starter strains. They can also be applied as bio-protective cultures in non-fermented foods, provided that they do not have adverse effects on the food. For application in foods stored under refrigeration, a desirable property would be the capacity to produce bacteriocin at low temperature. In some cases, a low bacteriocin production during refrigeration storage can be compensated with the simultaneous addition of a bacteriocin concentrate together with producer culture.

3.2 Application of LAB Bacteriocins as Part of Hurdle Technology

The efficacy of bacteriocins can improve considerably when applied in combination with other antimicrobials or barriers. As a matter of fact, food preservation most often relies on application of several barriers that restrict the survival and proliferation of microorganisms. These include treatments for inactivation of microorganisms in the raw materials, during processing, or in the finished product, acidification, addition of preservatives, and/or modification of atmosphere composition among others. The concept of hurdle technology (Leistner 2000) is based on the combination of different barriers acting in different ways on microbial cells, so that the cells have to activate different repair and adaptation mechanisms in order to survive and/or proliferate under the imposed selective conditions. Under such varied selective pressure, the cells will die as a consequence of energy exhaustion and failure to repair cell damages. Since most bacteriocins act on the bacterial cytoplasmic membrane, they interfere with the generation of energy required to repair bacterial cell damage. At the same time, cell damage or metabolic constraints imposed by other hurdles reduce the natural defense mechanisms of bacteria, making them susceptible at low bacteriocin concentrations that would not be lethal to intact cells. Some of the hurdles may also destabilize bacterial cell structures such as the outer membrane of Gram-negative bacteria. The outer membrane acts as a selective permeability barrier that retains bacteriocins and other antimicrobial substances. When this barrier is destabilized, bacteriocins (as well as other antimicrobials) can diffuse much better and reach the bacterial cytoplasmic membrane, where they act specifically.

The scientific literature is full of examples where bacteriocins have been tested in different food systems as part of hurdle technology, with several purposes (Gálvez et al. 2007, 2008): (1) enhancing the efficacy of bacteriocins as well as that of treatments; (2) reducing the required bacteriocin concentration; (3) broadening the spectrum of antimicrobial treatments (for example, to Gram-negative bacteria); (4) improving the inactivation of bacterial endospores; (5) as an additional barrier to proliferation of sublethally-injured cells as well as intact cells and endospores surviving treatments; and (6) as an additional barrier against post-process contamination. Specific examples of these applications will be discussed in the following chapters dealing with application of bacteriocins in different food systems. Notwithstanding, a summary of combined treatments and their reported effects is presented in Table 3.1.

Table 3.1 Application of bacteriocins in combined treatments

Antimicrobial treatment	Reported effects	Reference(s)
CO_2 atmosphere	Synergism with bacteriocins against foodborne pathogens (nisin, pediocin PA-1/Ach)	Nilsson et al. 2000; Szabo and Cahill 1999
Nitrates, nitrites	Enhanced antibacterial activity of nisin in meats against C. botulinum, Ln. mesenteroides and L. monocytogenes	Rayman et al. 1983; Gill and Holley 2003
Organic acids and salts	Improve bacteriocin solubility and increase sensitivity of target bacterial cells in different food substrates (nisin, pediocin PA-1/AcH, enterocin AS-48, or lacticin 3147…)	Scannell et al. 1997, 2000; Uhart et al. 2004; Cobo Molinos et al. 2005
	Sensitize Gram-negative bacteria due to their chelating activity	Scannell et al. 1997
Chelating agents (EDTA, polyphosphates)	Destabilization of the outer bacterial cell membrane in Gram-negative bacteria and improving cell permeability to bacteriocins	Stevens et al. 1991; Vaara 1992; Helander et al. 1997
	Sequestration of essential mineral nutrients and indirect sensitization to bacteriocins on Gram-positive bacteria	Gill and Holley 2000, 2003
Essential oils and their bioactive components	Greater inactivation of Gram-positive bacteria and sensitization of Gram-negative bacteria to bacteriocins	Pol et al. 2001a; Yuste and Fung 2004; Grande et al. 2007; Cobo Molinos et al. 2009
Other antimicrobials of chemical or proteinaceous nature[a]	Increased inactivation of Gram-positive and/or Gram-negative bacteria	Mulet-Powell et al. 1998; Gill and Holley 2000; Nattress and Baker 2003; Branen and Davidson 2004; Boussouel et al. 2000; Naghmouchi et al. 2010; Lüders et al. 2003
CO_2 atmosphere	Inhibition of strictly aerobic bacteria (complementary action with bacteriocins)	Economou et al. 2009
Heat	Increased inactivation of Gram-positive bacteria	Budu-Amoako et al. 1999; Ananou et al. 2004
	Perturbation of the bacterial outer cell membrane and sensitization of Gram-negative bacteria to bacteriocins	Kalchayanand et al. 1992; Boziaris et al. 1998; Ananou et al. 2005; Bakes et al. 2004
	Activation of endospore germination and reduced heat resistance of bacterial endospores	Wandling et al. 1999; Grande et al. 2006
	Additional barrier against endospores after heat treatments	Wandling et al. 1999; Grande et al. 2006

High intensity pulsed electric fields	Synergistic to additive effects on Gram-positive bacteria	Calderón-Miranda et al. 1999; Sobrino-Lopez and Martín Belloso 2006; Pol et al. 2001b
	Sensitization of Gram–negative bacteria, thus improving microbial inactivation in the combined treatments	Liang et al. 2002; Terebiznik et al. 2000; Martínez-Viedma et al. 2008
	Additional barrier against proliferation of survivors and sublethally-injured cells during the product shelf life	Martínez-Viedma et al. 2009
High hydrostatic pressure	Improved inactivation of Gram-positive bacteria	Ananou et al. 2010a; Ponce et al. 1998; Capellas et al. 2000; Arqués et al. 2005; López-Pedemonte et al. 2003
	Sensitization of Gram–negative bacteria, thus improving microbial inactivation in the combined treatments	Ponce et al. 1998; García-Graells et al. 1999; Black et al. 2005; Masschalck et al. 2001
	Additional barrier against proliferation of survivors and sublethally-injured cells during the product shelf life	López-Pedemonte et al. 2003; Arqués et al. 2005; Marcos et al. 2008
High pressure homogenization	Improved inactivation of L. monocytogenes in carrot juice	Pathanibul et al. 2009
Irradiation	Improved inactivation of L. monocytogenes, and protection against post-process contamination of the food	Chen et al. 2004
Pulsed light	Improved inactivation of L. monocytogenes, and protection against post-process contamination of the food	Uesugi and Moraru 2009

[a]lysozyme, lactoferrin, ovalbumin, lactoperoxidase system, and other antimicrobial peptides

References

Ananou S, Gálvez A, Martínez-Bueno M et al (2005) Synergistic effect of enterocin AS-48 in combination with outer membrane permeabilizing treatments against *Escherichia coli* O157:H7. J Appl Microbiol 99:1364–1372

Ananou S, Garriga M, Joffré A et al (2010a) Combined effect of enterocin AS-48 and high hydrostatic pressure treatment to control food-borne pathogens in slightly fermented sausages. Meat Sci 84:594–600

Ananou S, Muñoz A, Martínez-Bueno M et al (2010b) Evaluation of an enterocin AS-48 enriched bioactive powder obtained by spray drying. Food Microbiol 27:58–63

Ananou S, Valdivia E, Martínez Bueno M et al (2004) Effect of combined physico-chemical preservatives on enterocin AS-48 activity against the enterotoxigenic *Staphylococcus aureus* CECT 976 strain. J Appl Microbiol 97:48–56

Arqués JL, Rodríguez E, Gaya P et al (2005) Effect of combinations of high-pressure treatment and bacteriocin-producing lactic acid bacteria on the survival of *Listeria monocytogenes* in raw milk cheese. Int Dairy J 15:893–900

Bakes SH, Kitis FYE, Quattlebaum RG et al (2004) Sensitization of Gram-negative and Gram-positive bacteria to jenseniin G by sublethal injury. J Food Prot 67:1009–1013

Black EP, Kelly AL, Fitzgerald GF (2005) The combined effect of high pressure and nisin on inactivation of microorganisms in milk. Innov Food Sci Emerg Technol 6:286–292

Boussouel N, Mathieu F, Revol-Junelles AM et al (2000) Effects of combinations of lactoperoxidase system and nisin on the behaviour of *Listeria monocytogenes* ATCC 15313 in skim milk. Int J Food Microbiol 61:169–175

Boziaris IS, Humpheson L, Adams MR (1998) Effect of nisin on heat injury and inactivation of *Salmonella enteritidis* PT4. Int J Food Microbiol 43:7–13

Branen JK, Davidson PM (2004) Enhancement of nisin, lysozyme, and monolaurin antimicrobial activities by ethylenediaminetetraacetic acid and lactoferrin. Int J Food Microbiol 90:63–74

Budu-Amoako E, Ablett RF, Harris J et al (1999) Combined effect of nisin and moderate heat on destruction of *Listeria monocytogenes* in cold-pack lobster meat. J Food Prot 62:46–50

Calderón-Miranda ML, Barbosa-Canovas GV, Swanson BG (1999) Inactivation of *Listeria innocua* in skim milk by pulsed electric fields and nisin. Int J Food Microbiol 51:19–30

Capellas M, Mor-Mur M, Gervilla R et al (2000) Effect of high pressure combined with mild heat or nisin on inoculated bacteria and mesophiles of goats' milk fresh cheese. Food Microbiol 17:633–641

Chen CM, Sebranek JG, Dickson JS et al (2004) Combining pediocin with post-packaging irradiation for control of *Listeria monocytogenes* on frankfurters. J Food Prot 67:1866–1875

Cobo Molinos A, Abriouel H, Ben Omar N et al (2009) Enhanced bactericidal activity of enterocin AS-48 in combination with essential oils, natural bioactive compounds, and chemical preservatives against *Listeria monocytogenes* in ready-to-eat salads. Food Chem Toxicol 47:2216–2223

Cobo Molinos A, Abriouel H, Ben Omar N et al (2005) Effect of immersion solutions containing enterocin AS-48 on *Listeria monocytogenes* in vegetable foods. Appl Environ Microbiol 71:7781–7787

Deegan LH, Cotter PD, Hill C et al (2006) Bacteriocins: biological tools for bio-preservation and shelf-life extension. Int Dairy J 16:1058–1071

Economou T, Pournis N, Ntzimani A et al (2009) Nisin-EDTA treatments and modified atmosphere packaging to increase fresh chicken meat shelf-life. Food Chemist 114:1470–1476

Gálvez A, Abriouel H, López RL et al (2007) Bacteriocin-based strategies for food biopreservation. Int J Food Microbiol 120:51–70

Gálvez A, Lucas R, Abriouel H et al (2008) Application of bacteriocins in the control of foodborne pathogenic and spoilage bacteria. Crit Rev Biotechnol 28:125–152

Garcia-Graells C, Masschalck B, Michiels CW (1999) Inactivation of *Escherichia coli* in milk by high-hydrostatic-pressure treatment in combination with antimicrobial peptides. J Food Prot 62:1248–1254

Gill AO, Holley RA (2000) Surface application of lysozyme, nisin, and EDTA to inhibit spoilage and pathogenic bacteria on ham and bologna. J Food Prot 63:1338–1346

Gill AO, Holley RA (2003) Interactive inhibition of meat spoilage and pathogenic bacteria by lysozyme, nisin and EDTA in the presence of nitrite and sodium chloride at 24 °C. Int J Food Microbiol 80:251–259

Grande MJ, Lucas R, Abriouel H et al (2007) Treatment of vegetable sauces with enterocin AS-48 alone or in combination with phenolic compounds to inhibit proliferation of *Staphylococcus aureus*. J Food Prot 70:405–411

Grande MJ, Lucas R, Abriouel H et al (2006) Inhibition of toxicogenic *Bacillus cereus* in rice-based foods by enterocin AS-48. Int J Food Microbiol 106:185–194

Helander IM, von Wright A, Mattila-Sandholm TM (1997) Potential of lactic acid bacteria and novel antimicrobials against Gram-negative bacteria. Trends Food Sci Technol 8:146–150

Kalchayanand N, Hanlin MB, Ray B (1992) Sublethal injury makes Gram-negative and resistant Gram-positive bacteria sensitive to the bacteriocins, pediocin AcH and nisin. Lett Appl Microbiol 16:239–243

Leistner L (2000) Basic aspects of food preservation by hurdle technology. Int J Food Microbiol 55:181–186

Liang Z, Mittal GS, Griffiths MW (2002) Inactivation of *Salmonella*Typhimurium in orange juice containing antimicrobial agents by pulsed electric field. J Food Prot 65:1081–1087

López-Pedemonte T, Roig-Sagués AX, Trujillo AJ et al (2003) Inactivation of spores of *Bacillus cereus* in cheese by high hydrostatic pressure with the addition of nisin or lysozyme. J Dairy Sci 86:3075–3081

Lüders T, Birkemo GA, Fimland G et al (2003) Strong synergy between a eukaryotic antimicrobial peptide and bacteriocins from lactic acid bacteria. Appl Environ Microbiol 69:1797–1799

Marcos B, Aymerich T, Monfort JM et al (2008) High-pressure processing and antimicrobial bio-degradable packaging to control *Listeria monocytogenes* during storage of cooked ham. Food Microbiol 25:177–182

Martínez-Viedma P, Abriouel H, Sobrino A et al (2009) Effect of enterocin AS-48 in combination with High-Intensity Pulsed-Electric Field treatment against the spoilage bacterium *Lactobacillus diolivorans* in apple juice. Food Microbiol 26:491–496

Martínez-Viedma P, Sobrino A, Omar B et al (2008) Enhanced bactericidal effect of High-Intensity Pulsed-Electric Field treatment in combination with enterocin AS-48 against *Salmonella enterica* in apple juice. Int J Microbiol 128:244–249

Masschalck B, Van Houdt R, Michiels CW (2001) High pressure increases bactericidal activity and spectrum of lactoferrin, lactoferricin and nisin. Int J Food Microbiol 64:325–332

Morgan SM, Galvin M, Kelly J et al (1999) Development of a lacticin 3147-enriched whey powder with inhibitory activity against foodborne pathogens. J Food Prot 62:1011–1016

Mulet-Powell N, Lacoste-Armynot AM, Vinas M et al (1998) Interactions between pairs of bacteriocins from lactic bacteria. J Food Prot 61:1210–1212

Naghmouchi K, Drider D, Baah J et al (2010) Nisin A and polymyxin B as synergistic inhibitors of Gram-positive and Gram-negative bacteria. Probiot Antim Prot 2:98–103

Nattress FM, Baker LP (2003) Effects of treatment with lysozyme and nisin on the microflora and sensory properties of commercial pork. Int J Food Microbiol 85:259–267

Nilsson L, Chen Y, Chikindas ML et al (2000) Carbon dioxide and nisin act synergistically on *Listeria monocytogenes*. Appl Environ Microbiol 66:769–774

O'Mahony T, Rekhif N, Cavadini C et al (2001) The application of a fermented food ingredient containing 'variacin,' a novel antimicrobial produced by *Kocuria varians*, to control the growth of *Bacillus cereus* in chilled dairy products. J Appl Microbiol 90:106–114

Pathanibul P, Taylor TM, Davidson PM et al (2009) Inactivation of *Escherichia coli* and *Listeria innocua* in apple and carrot juices using high pressure homogenization and nisin. Int J Food Microbiol 129:316–320

Pol IE, Mastwijk HC, Slump RA et al (2001a) Influence of food matrix on inactivation of *Bacillus cereus* by combinations of nisin, pulsed electric field treatment, and carvacrol. J Food Prot 64:1012–1018

Pol IE, van Arendonk WG, Mastwijk HC et al (2001b) Sensitivities of germinating spores and carvacrol-adapted vegetative cells and spores of *Bacillus cereus* to nisin and pulsed-electric-field treatment. Appl Environ Microbiol 67:1693–1699

Ponce E, Pla R, Sendra E et al (1998) Combined effect of nisin and high hydrostatic pressure on destruction of *Listeria innocua* and *Escherichia coli* in liquid whole egg. Int J Food Microbiol 43:15–19

Rayman K, Malik N, Hurst A (1983) Failure of nisin to inhibit outgrowth of *Clostridium botulinum* in a model cured meat system. Appl Environ Microbiol 46:1450–1452

Robertson A, Tirado C, Lobstein T et al (eds) (2004) Food and health in Europe: a new basis for action. WHO Regional Publications, European Series, No. 96, Geneva

Ross RP, Morgan S, Hill C (2002) Preservation and fermentation: past, present and future. Int J Food Microbiol 79:3–16

Scannell AG, Ross RP, Hill C et al (2000) An effective lacticin biopreservative in fresh pork sausage. J Food Prot 63:370–375

Scannell AGM, Hill C, Buckley DJ et al (1997) Determination of the influence of organic acids and nisin on shelf-life and microbiological safety aspects of fresh pork sausage. J Appl Microbiol 83:407–412

Sobrino-Lopez A, Martin Belloso O (2006) Enhancing inactivation of *Staphylococcus aureus* in skim milk by combining high-intensity pulsed electric fields and nisin. J Food Prot 69:345–353

Stevens KA, Sheldon BW, Klapes NA et al (1991) Nisin treatment for inactivation of *Salmonella* species and other gram-negative bacteria. Appl Environ Microbiol 57:3613–3615

Szabo EA, Cahill ME (1999) Nisin and ALTA™ 2341 inhibit the growth of *Listeria monocytogenes* on smoked salmon packaged under vacuum or 100 % CO_2. Lett Appl Microbiol 28:373–377

Terebiznik MR, Jagus RJ, Cerrutti P et al (2000) Combined effect of nisin and pulsed electric fields on the inactivation of *Escherichia coli*. J Food Prot 63:741–746

Thomas LV, Clarkson MR, Delves-Broughton J (2000) Nisin. In: Naidu AS (ed) Natural food antimicrobial systems. CRC-Press, Boca Raton, FL, pp 463–524

Uesugi AR, Moraru CI (2009) Reduction of *Listeria* on ready-to-eat sausages after exposure to a combination of pulsed light and Nisin. J Food Prot 72:347–353

Uhart M, Ravishankar S, Maks ND (2004) Control of *Listeria monocytogenes* with combined antimicrobials on beef franks stored at 4 °C. J Food Prot 67:2296–2301

Vaara M (1992) Agents that increase the permeability of the outer membrane. Microbiol Rev 56:395–411

Wandling LR, Sheldon BW, Foegeding PM (1999) Nisin in milk sensitizes *Bacillus* spores to heat and prevents recovery of survivors. J Food Prot 62:492–498

Yuste J, Fung DY (2004) Inactivation of *Salmonella* Typhimurium and *Escherichia coli* O157:H7 in apple juice by a combination of nisin and cinnamon. J Food Prot 67:371–377

Chapter 4
Biopreservation of Meats and Meat Products

4.1 Application of Bacteriocin Preparations

4.1.1 Raw Meats

The microbial populations most frequently associated with the meat environment are known to primarily belong to the groups Enterobacteriaceae, lactic acid bacteria (LAB), *Brochothrix thermosphacta*, and pseudomonads (Borch et al. 1996; Labadie 1999; Nychas et al. 2008). Microbial metabolism of meat during growth results in microbial spoilage, with the development of offodors which make the product undesirable for human consumption (Jackson et al. 1997). Also, pathogenic bacteria initially present at low concentrations may grow during meat spoilage may proliferate during refrigeration storage, especially *Listeria monocytogenes*.

In raw meats, bacteriocins have been tested alone or in combination with other hurdles for carcass decontamination and/or to inhibit bacterial growth on stored fresh meats (Table 4.1). Washing, spraying or dipping with bacteriocin solutions have been tested alone or in combination with other antimicrobials to potentiate bacteriocin activity. In order to increase the efficacy of treatments and/or avoid cross contamination, raw meats are chilled, packaged under different atmospheric conditions such as vacuum packaging, MAP, or active packaging with O_2 scavengers or CO_2 generating systems (Coma 2008; McMillin 2008). Additional combinations such as low dose irradiation, UV surface decontamination or HHP have been proposed (Aymerich et al. 2008). All these processing treatments have selective effects the initial microbiota, and may act in synergy with bacteriocins to increase the product safety and shelf life. Although raw meat products are further processed prior to consumption by treatments that usually destroy pathogenic bacteria, they can be a considerable source of cross contamination. Growth of toxin-producing bacteria in raw materials (such as minced meats) should also be controlled, especially for heat-stable toxins.

© The Author(s) 2014
A. Gálvez et al., *Food Biopreservation*, SpringerBriefs in Food, Health, and Nutrition, DOI 10.1007/978-1-4939-2029-7_4

Table 4.1 Examples of applications of bacteriocin preparations in meat and poultry products

Bacteriocin preparations	Effect(s)	Reference(s)
Raw meats		
Nisin combinations (organic acids, chelators, lysozyme, vacuum packaging, MAP)	Decontamination of raw meat surfaces before processing	Thomas et al. 2000
Nisin activated film with EDTA	Inhibition of LAB, carnobacteria and *B. thermosphacta* and reduction of *Enterobacteriaceae* load on beef cuts	Ercolini et al. 2010
Pediocins	Anti-listeria protection by pediocins in raw meats	Rodríguez et al. 2002
Pentocin 31-1	Reduction of growth of *Listeria* and *Pseudomonas* and total volatile basic nitrogen production in chill-stored tray-packaged pork meat	Zhang et al. 2010
RTE meats		
Nisin activated films	Increased inactivation of *L. monocytogenes* in several vacuum-packaged products	Aymerich et al. 2008
Nisin in combination with HHP	Increased inactivation of *E. coli* and staphylococci in cooked ham, avoiding regrowth of *E. coli* and slime-forming bacteria	Garriga et al. 2002
Nisin and pulsed light	Application of a Nisaplin dip followed by exposure to pulsed light reduced the population of *L. innocua* on sausages	Uesugi and Moraru 2009
Nisin-pectin film, in combination with low-dose irradiation	Increased microbial inactivation of *L. monocytogenes* on RTE turkey meat and inhibition of survivor proliferation during storage	Jin et al. 2009
Pediocin in combination with post-packaging irradiation or thermal treatment	Effective combination with to control *L. monocytogenes* on frankfurters	Chen et al. 2004a, b
Enterocin alginate film, in combination with HHP	Prevention of *L. monocytogenes* regrowth in the treated cooked ham during cold storage as well as during cold chain break	Marcos et al. 2008a, b

At present, there is a great body of research data concerning bacteriocin trials on raw meats, many of them dealing with nisin. Nisin has been widely tested for preservation of raw meats (Thomas et al. 2000). However, the application of nisin in meats has several drawbacks such as its poor solubility, interaction with phospholipids and antagonism by glutathione (Thomas et al. 2000; Stergiou et al. 2006). Nevertheless, positive results have been reported for surface decontamination of raw meats before processing and packaging, in which antimicrobial activity was potentiated by combination with other antimicrobials or hurdles such as organic acids, chelators, vacuum packaging, or MAP. Representative examples reported on vacuum-packaged beef are the reduction in the numbers of *Listeria innocua* and *B. thermosphacta* after nisin treatment (Cutter and Siragusa 1996a) or the inhibition of *L. monocytogenes* and *Escherichia coli* O157:H7 after treatment with nisin and

EDTA (Zhang and Mustapha 1999). Similarly, dipping in solutions containing combinations of lactic or polylactic acids and nisin reduced the microbial load of meats before processing and afforded an extended shelf-life in vacuum-packaged fresh meat (Ariyapitipun et al. 1999, 2000; Barboza de Martinez et al. 2002), and treatment with a combination of nisin and lysozyme effectively inhibited *B. thermosphacta* and LAB in vacuum-packaged pork (Nattress et al. 2001; Nattress and Baker 2003). Other reports indicated that, under MAP, nisin was able to completely inhibit growth of *L. monocytogenes* in pork (Fang and Lin 1994a, b).

In raw poultry meats, application of antimicrobial treatments using nisin and EDTA to in combination with MAP or vacuum packaging (VP) reduced total aerobic plate counts and increased the product shelf-life by a minimum of 4 days when packaged under aerobic conditions and a maximum of 9 days when vacuum packaged (Cosby et al. 1999). The use of MAP (65 % CO_2, 30 % N_2, 5 % O_2) in combination with nisin–EDTA antimicrobial treatments affected the populations of mesophilic bacteria, *Pseudomonas* sp., *B. thermosphacta*, lactic acid bacteria and *Enterobacteriaceae*, and resulted in an organoleptic extension of refrigerated, fresh chicken meat for up to 14 days, decreasing the formation of volatile amines, trimethylamine nitrogen and total volatile nitrogen (Economou et al. 2009).

Treatment of raw meats and poultry meats with pediocins (especially pediocin PA-1/Ach) singly or in combination with other hurdles can inhibit or delay growth of spoilage Gram-positive bacteria (such as *B. thermosphacta*) and/or reduce *L. monocytogenes* populations (Rodríguez et al. 2002; Nieto-Lozano et al. 2006; Kalchayanand 1990; Nielsen et al. 1990; Motlagh et al. 1992; Degnan et al. 1993; Schlyter et al. 1993: Taalat et al. 1993; Goff et al. 1996; Murray and Richard 1997). For example, treatment of raw meat surfaces with 500, 1,000 or 5,000 bacteriocin units/ml (BU/ml) reduced the counts of inoculated *L. monocytogenes* after storage at 15 °C during 72 h by 1, 2 or 3 log cycles, and treatment with 1,000 or 5,000 BU/ml reduced its viable counts by 2.5 or 3.5 log cycles, respectively, after storage at 4 °C during 21 days compared to the control not treated with bacteriocin. The same bacteriocin treatments exerted a bacteriostatic effect on *Clostridium perfringens* (Nieto-Lozano et al. 2006). In poultry meats, treatment with pediocin PA-1/Ach adsorbed to heat killed *Pediococcus acidilactici* cells was very effective in the control of *L. monocytogenes* in refrigerated chicken meat (Goff et al. 1996).

Other bacteriocins such as sakacins, carnobacteriocins, bifidocins, lactocins, lactococcins, enterocins or pentocins have shown variable inhibitory effects against spoilage or pathogenic bacteria in raw meats or poultry meats (Aymerich et al. 2000, 2008; Galvez et al. 2008). In chicken breasts, addition of enterocins A and B produced by the meat isolate *Enterococcus faecium* CTC492 (4,800 AU/cm^2) reduce the population of *Listeria* to 3.6 MNP/cm^2 during incubation at 7 °C (Aymerich et al. 2000). In vacuum-packed chicken cuts stored under refrigeration, treatment with sakacin-P caused strong inhibition of *L. monocytogenes* (Katla et al. 2002). Addition of bifidocin B (from *Bifidobacterium bifidum*) and lactococcin R (produced by *Lactococcus lactis* subsp. *cremoris*) to irradiated raw chicken breast inhibited the growth of *L. monocytogenes* or *Bacillus cereus* for 3–4 weeks at 5–8 °C or 6–12 h at 22–25 °C (Yildirim et al. 2007). Another study reported that application of pentocin 31-1 (produced by a *Lactobacillus pentosus* strain isolated from the traditional Chinese fermented Xuan-Wei Ham) in chill-stored non-vacuum tray-packaged pork meat substantially reduced the growth of *Listeria* and *Pseudomonas* as well as the total volatile basic nitrogen (measured as an indicator of meat spoilage) during cold storage compared with the untreated control (Zhang et al. 2010).

One attractive approach to optimize the activity of bacteriocins in raw meats has been immobilisation in substrates (such as beads, liposomes, coatings or films). Nisin (alone or in combinations with citric acid, EDTA, and Tween 80) incorporated in a variety of substrates (such as calcium alginate gels, agar coatings, palmitoylated alginate-based films, polyvinyl chloride, LDPE, or nylon) showed strong inhibition of bacteria such as *L. monocytogenes*, *B. thermosphacta*, *Staphylococcus aureus*, or *Salmonella* Typhimurium on refrigerated raw meats (Chen and Hoover 2003; Aymerich et al. 2008; Gálvez et al. 2007, 2008). This approach decreases the impact of interaction with food components and enzyme inactivation of bacteriocin activity, and also decreases the amount of bacteriocin required for inhibition of target bacteria (Quintavalla and Vicini 2002). In raw meats and poultry samples packaged in bags coated with pediocin powder, the pediocin completely inhibited growth of inoculated *L. monocytogenes* through 12 weeks storage at 4 °C (Ming et al. 1997). Application of nisin immobilized in calcium alginate gel on beef carcass tissues completely suppressed *B. thermosphacta* (Cutter and Siragusa 1996b), and low density polyethylene films containing nisin prevented carcass contamination by this bacterium (Siragusa et al. 1999). Nisin bound to activated alginate beads or in a palmitoylated alginate-based film (to avoid nisin degradation) reduced the viable counts of *S. aureus* in ground beef and on sliced beef meat, respectively (Millette et al. 2007). Treatment of fresh poultry with agar coatings containing nisin achieved substantial reductions in *S.* Typhimurium growth after storage at 4 °C for 96 h (Natrajan and Sheldon 1995). The efficiency of numerous films formulation based on polyvinyl chloride, LDPE, nylon, calcium-alginate or agar containing nisin (in combinations with citric acid, EDTA, and Tween 80) to inhibit the antibiotic resistant *S.* Typhimurium on poultry drumstick skin was demonstrated by the same researchers (Natrajan and Sheldon 2000a, b). Combinations of nisin with citric acid, EDTA and Tween-80 also led to a 4–5 log reduction of psychotrophic aerobes during 72 h of storage. In another study, Ercolini et al. (2010) tested a nisin activated plastic antimicrobial packaging (developed by using a nisin, HCl and EDTA solution) on beef cuts stored at 1 °C. The combination of chill temperature and antimicrobial packaging proved to be effective in enhancing the microbiological quality of beef cuts by inhibiting LAB, carnobacteria and *B. thermosphacta* in the early stages of storage and by reducing the loads of *Enterobacteriaceae*, without affecting the species diversity according to PCR-DGGE fingerprints of DNA extracted from the treated meat cuts (Ercolini et al. 2010). Also, plastic bags activated at their internal face with a nisin-EDTA solution were used for vacuum-packaging of beef chops (Ferrocino et al. 2013). During storage in the activated films at 1 °C, *B. thermosphacta* was unable to grow for the whole storage time (46 days), while the levels of *Carnobacterium* spp. were below the detection limit for the first 9 days and reached levels below 5 log CFU/cm^2 after 46 days. The antimicrobial packaging had no effect on *Enterobacteriaceae* or *Pseudomonas* spp., with final populations of about 4 log CFU/cm². Nevertheless, the active packaging reduced the release of volatile metabolites in the headspace of beef with a probable positive impact on meat quality. Recycling of industrial wastes into useful products is a growing trend not only in the food industry but in many other fields as well. In a recent and innovative study, a novel poly(lactic acid)/sawdust particle biocomposite film with anti-listeria activity was developed by incorporation of pediocin PA-1/AcH (Woraprayote et al. 2013). It was reported that sawdust particle played an important role in embedding pediocin into the hydrophobic PLA film. Application of the activated film as a food-contact antimicrobial packaging on raw sliced pork efficiently inhibited *L. monocytogenes* during chill storage.

The bacteriocin 32Y (from *Lactobacillus curvatus* 32Y) was used to develop an industrially produced activated plastic film (Mauriello et al. 2004). In experiments of food packaging with pork steak and ground beef (simulating hamburgers) contaminated by *L. monocytogenes* V7, highest antimicrobial activity was observed after 24 h at 4 °C, with a decrease of about 1 log of the *L. monocytogenes* population (Mauriello et al. 2004). The lactocins 705 and AL705 are produced by *L. curvatus* CRL705. Lactocin 705 has antagonist effect against LAB and *B. thermosphacta*, while AL705 is active against *Listeria* species (Castellano and Vignolo 2006). Both bacteriocins retained antimicrobial activity when included in polymer matrices such as LDPE (Blanco et al. 2008, 2012) and gluten (Blanco Massani et al. 2014). In trials with *L. curvatus* CRL705 immobilized bacteriocins, a bacteriostatic effect against *L. innocua* 7 was observed in both synthetic (Cryovac films) and gluten activated packages until the fourth week of storage (Blanco Massani et al. 2014).

The process operations for manufacture of minced meats facilitate inoculation of contaminating bacteria in the meat batter. Therefore, the presence and multiplication of foodborne pathogens in minced meats should be controlled. One study showed that the single addition of nisin extended the lag phase of *L. monocytogenes* inoculated into minced buffalo meat (Pawar et al. 2000). In minced meats, the combination of bacteriocins with plant essential oils at levels where they would not impart undesirable flavour is being considered as a way to increase inactivation of *L. monocytogenes* and inhibition of *Salmonella* Enteritidis (Solomakos et al. 2008; Govaris et al. 2010). Antilisterial activity of nisin in minced beef increased greatly in combination with thyme essential oil. The combination of essential oil at 0.6 % with nisin at 1,000 IU/g decreased the population of *L. monocytogenes* below the official limit set by European Union during storage at 4 °C for at least 12 days (Solomakos et al. 2008). At that concentration, the thyme oil did not impart undesirable flavour. Promising results have also been reported on inhibition of *S.* Enteritidis in sheep minced meat by a combination of nisin and oregano essential oil (Govaris et al. 2010), while the single treatment of minced sheep meat with nisin at 500 or 1,000 IU/g had no activity against *S.* Enteritidis. The combination of the oregano essential oil at 0.6 % with nisin at 500 IU/g showed stronger antimicrobial activity against *S.* Enteritidis than the single oregano essential oil at 0.6 % but lower than the combination with nisin at 1,000 IU/g (Govaris et al. 2010). Best results were reported for the combinations of oregano essential oil at 0.9 % with nisin at 500 or

1,000 IU/g, which showed a bactericidal effect against the pathogen. The inhibitory effects were higher in samples stored at 10 °C compared to 4 °C. This could be a draw-back for cold-stored meats, but at the same time could be an advantage under episodes of cold chain break and temperature abuse. Regarding the effects of other bacteriocins in minced meats, addition of a partially-purified plantaricin preparation from *Lactobacillus plantarum* UG1 rapidly reduced the population of *L. monocytogenes* below detectable levels in minced meat stored at 8 °C, (Enan et al. 2002), and the addition of a freeze-dried whey fermentate from *C. piscicola* (containing piscicocin CS526) to a ground mixture of beef and pork meat reduced the population of *L. monocytogenes* below detectable levels for at least 4 days at 12 °C and for up to 25 days at 4 °C (Azuma et al. 2007). Furthermore, in minced pork treated with a preparation of enterocins A and B (1,600 AU/g) from *E. faecium* CTC492, the levels of listeria were reduced below 3 MNP/g after 6 days of incubation at 7 °C while the untreated control increased from 5 MNP/g to 48 CFU/g (Aymerich et al. 2000).

4.1.2   Semi-processed and Cooked Meats

Cooked meat products are widely consumed ready-to-eat (RTE) foods. They may consist of whole primary meat pieces, but usually they are made by grinding and mixing secondary meats, fat, animal organs, or blood with other ingredients, followed by stuffing/molding and cooking. The cooking process inactivates natural microbiota, paving the way for growth of post-process contaminants. The pH values of most cooked meat products are compatible with growth of pathogenic and spoilage bacteria, which can proliferate at refrigeration temperatures during the product shelf life. Some of these meats may also undergo further processing such as slicing, peeling, and packaging, which increase the risks for cross-contamination (Murphy et al. 2005). For these reasons, there has been a great interest in the application of bacteriocins (mainly pediocin and nisin) as hurdles against spoilage bacteria and pathogens (mainly *L. monocytogenes*). The main approaches tested are based on addition of bacteriocin preparations to the meat slurries before the heating process, surface application of the bacteriocins before packaging, or application of films or coatings dosed with bacteriocins. The possibility of adding bacteriocins in the meat before the cooking process due to their thermotolerance is of great interest.

Strains of LAB (mainly *Lactobacillus* and *Leuconostoc*) are the major group of spoilage bacteria developing on various types of vacuum-packed meats, where they produce typical sensory changes such as souring, gas, SH_2 and slime (Korkeala et al. 1988; Björkroth and Korkeala 1997). In one study using sakacin K, nisin and enterocins, the results obtained clearly depended on the bacteriocin and the target bacteria (Aymerich et al. 2002). Sakacin K and nisin were unable to prevent ropiness caused by *Lactobacillus sakei* CTC746 strain, but nisin was able to prevented ropiness caused by *Leuconostoc carnosum* CTC747 (Aymerich et al. 2002). Nisin was also the most effective bacteriocin on staphylococci, but did not prevent regrowth of *L. monocytogenes* (while enterocins, sakacin and pediocin did).

Ovotransferrin is the main component in the antimicrobial defense system of hens' egg. Antimicrobial activity of ovotransferrin is mainly due to its iron-binding capacity, but direct interactions with the bacterial surface also seem to play an important role in contributing to its inhibitory activity (Moon et al. 2011). Ovotransferrin, nisin, and their combinations had strong antilisterial activity in BHI broths. However, addition of ovotransferrin to frankfurters did not inhibit growth of *L. monocytogenes*. When nisin (1,000 IU/frankfurter) was applied, an early bactericidal effect followed by delayed growth was observed (Moon et al. 2011). However, no differences were reported in the antilisterial effect when the same nisin concentration was applied in combination with ovotransferrin (40 mg/frankfurter). The observed differences could be explained by the influence of factors such as interaction with food substrate or a higher iron content in meat.

Incorporation of nisin into bologna-type sausages during mixing of ingredients inhibited the growth of spoilage LAB during further storage at 8 °C of the resulting vacuum-packed sausages (Davies and Delves-Broughton 1999). The effectiveness of nisin against several bacteria (such as *B. thermosphacta*, *L. curvatus*, *Ln. mesenteroides*, *L. monocytogenes*, *Salmonella* sp. and *E. coli* O157:H7) in ham and/or bologna sausages increased in combination with lysozyme and EDTA (Gill and Holley 2000a, b). In fresh pork sausages, a combination of nisin and organic acids reduced the viable counts of *Salmonella* Kentucky and *S. aureus* (Scannell et al. 1997). The combination of sodium citrate or sodium lactate with nisin or lacticin 3147 was also reported to increase the inhibition of *Listeria* and *C. perfringens* in fresh pork sausages (Scannell et al. 2000a).

Pediocin activity was increased when added in combinations with sodium diacetate or sodium lactate against *L. monocytogenes* on frankfurters or *L. monocytogenes* and *Yersinia enterocolitica* on cooked poultry cuts stored under MAP at 3.5 °C (O'Sullivan et al. 2002; Chen and Hoover 2003; Aymerich et al. 2008). The antilisterial activity of pediocin in slurries prepared from ready-to-eat turkey breast meat increased greatly when tested in combination with diacetate, due to synergistic effects between the two antimicrobials (Schlyter et al. 1993). When commercial beef franks were dipped for 5 min in three antimicrobial solutions: pediocin (6,000 AU), 3 % sodium diacetate and 6 % sodium lactate combined, and a combination of the three antimicrobials, reductions of *L. monocytogenes* populations ranged between 1 and 1.5 log units and 1.5–2.5 log units after 2 and 3 weeks of storage, respectively, at 4 °C (Uhart et al. 2004). These results indicated that the use of combined antimicrobial solutions for dipping treatments is more effective at inhibiting *L. monocytogenes* than treatments using antimicrobials such as pediocin separately (Uhart et al. 2004). In another study, the effects and interactions of temperature (56.3–60 °C), added sodium lactate (0–4.8 %) and sodium diacetate (0–0.25 %) and dipping in pediocin (0–10,000 AU) on *L. monocytogenes* in bologna were studied by Maks et al. (2010). Combination treatments increased or decreased *D*-values, depending on the temperature. Pediocin (2,500 and 5,000 AU) and heat decreased *D*-values, but pediocin exhibited a protective effect at higher concentrations ($\geq$7,500 AU). The results showed that interactions between additives in formulations can vary at different temperatures/concentrations, thereby affecting thermal inactivation of foodborne pathogens in meat products.

Enterocins have also been tested in cooked meat products. Addition of a partially-purified preparation of enterocins A and B (4,800 AU/g) reduced the numbers of *L. innocua* by 7.98 log cycles in cooked ham and by 9 log cycles in pork liver paté stored at 7 °C for 37 days (Aymerich et al. 2000). In vacuum packaged sliced cooked pork ham, added enterocins A and B (128 AU/g) inhibited the production of slime by *Lactobacillus sakei* CTC746 strain, but not by *Leuconostoc carnosum* CTC747 strain (Aymerich et al. 2002).

Results from studies on the synergistic activities of bacteriocins with other anti-microbials and on the effect of immobilized preparations or application of bacterio-cins by dipping solutions, together with the technical advances in the development of activated supports opened the doors for application of immobilized bactericin preparations or activated packagings containing cocktails of antimicrobial sub-stances on RTE meats (Coma 2008). Bacteriocin-activated films may be quite use-ful for cooked meat products, not only because they can prolong the product shelf life by decreasing the risks of spoilage and growth of pathogens from cross-contamination during processing, but also because the film itself acts as a barrier against external contamination of the processed product. Among the various kinds of edible coatings tested on vacuum-packaged products (hot dogs, frankfurters, or ham) best results have been reported for coatings containing nisin in combination with other antimicrobials under refrigeration storage.

Application of zein coatings containing nisin, sodium lactate, and sodium diace-tate completely eliminated *L. monocytogenes* on turkey frankfurters during refrig-eration storage (Lungu and Johnson 2005). In hot dogs that were vacuum-packaged in films coated with nisin, *L. monocytogenes* counts decreased during refrigeration storage (Franklin et al. 2004). Hot dogs were placed in control and nisin-containing pouches and inoculated with a five-strain *L. monocytogenes* cocktail (approximately 5 log CFU per package), vacuum sealed, and stored for intervals of 2 h and 7, 15, 21, 28, and 60 days at 4 °C. In hot dogs packaged in films coated with 2,500 IU/ml nisin solution, nisin significantly decreased ($P<0.05$) *L. monocytogenes* populations on the surface of hot dogs by greater than 2 log CFU per package throughout the 60-days study. However, *L. monocytogenes* populations still remained at approxi-mately 4 log CFU per package after 60 days of refrigerated storage (Franklin et al. 2004). This study reported similar results when using a cellulose-based coating solu-tion (based on methylcellulose/hydroxypropyl methylcellulose) containing nisin. However, in another study nisin-coated cellulose casings showed only moderate antilisterial activity in vacuum-sealed frankfurters, unless additional antimicrobials, such as potassium lactate and sodium diacetate, were employed (Luchansky and Call 2004). Nguyen et al. (2008) carried out similar experiments using an edible bacterial cellulose film containing nisin to control *L. monocytogenes* and total aero-bic bacteria on the surface of vacuum-packaged frankfurters. The frankfurters pack-aged in films activated with 2,500 IU/ml showed significantly lower counts of *L. monocytogenes* and total aerobic plate counts during refrigerated storage for 14 days as compared to the controls. The authors concluded that activated cellulose films had potential applicability as antimicrobial packaging films or inserts for pro-cessed meat products. Another study reported that polythene films activated with bacteriocin 32Y from *L. curvatus* were effective in reducing the population of listeria in vacuum-packaged frankfurters during storage at 4 °C (Ercolini et al. 2006). By using viable staining and fluorescence microscopy, the authors corroborated that the activated film caused an immediate reduction of live and appearance of dead cells just after 15 min from the packaging.

Another suggested application of nisin is the preservation of natural sausage casings. Casings derived from animal intestines can be one possible route for transmission of *C. perfringens* spores and other sulphite-reducing anaerobic spores, since the brining process of intestines does not inactivate bacterial endospores. In one study, it was shown that nisin was partly reversibly bound to casings and can reduce the outgrowth of *Clostridium sporogenes* spores in the model used by approximately 1 log cycle (Wijnker et al. 2011). This could open new possibilities to combat the entry of pathogens in the food chain.

In vacuum-packaged cooked ham, application of a gelatine coating gel containing a combination of lysozyme, nisin and EDTA in showed bactericidal activity for *B. thermosphacta*, *L. sakei*, *L. mesenteroides*, *L. monocytogenes* and *S. enterica* serovar Typhimurium (Gill and Holley 2000b). In sliced cooked ham packaged under MAP and stored at 4 °C, the inclusion of polyethylene/polyamide inserts coated with nisin (approximately 2,560 AU cm^2) reduced the levels of LAB, *Listeria innocua* and *Staphylococcus aureus*, and partially inhibited growth of total aerobic bacteria on the ham during storage (Scannell et al. 2000b). However, in ham steaks packaged in chitosan-coated plastic films containing 500 IU/cm^2 of nisin, the low bacteriocin concentration tested was ineffective in inhibiting *L. monocytogenes* (Ye et al. 2008).

Pediocin immobilization has also shown variable results. Encapsulation of pediocin AcH in liposomes enhanced its antimicrobial activity in meat slurries (Degnan and Luchansky 1992). However, in another study, when pediocin adsorbed to its heat-killed producer cells was used to treat sliced frankfurters before packaging, the number of *L. monocytogenes* decreased during 6 days of storage, but remained at constant levels for the remaining storage period (up to 21 days), indicating that the pediocin preparation was not efficient enough to kill all *L. monocytogenes* (Mattila et al. 2003). The efficacy of cellulose films containing pediocin PA-1/Ach (ALTA® 2351) against *L. innocua* and *Salmonella* sp. was tested on sliced ham packaged under vacuum and stored at 12 °C simulating abusive temperatures that can occur in supermarkets (Santiago-Silva et al. 2009). The antimicrobial films were more effective inhibiting growth of *L. innocua* (with a growth reduction of 2 log cycles compared to control treatment after 15 days of storage) than *Salmonella* (0.5 log cycle reduction in relation to control, after 12 days). However, the viable cell concentrations of the inoculated bacteria were not reduced for any of the treatments.

Films activated with enterocin 496K1 (from *Enterococcus casseliflavus* IM 416K1) and enterocins A and B have been tested in ready to eat meat products (Iseppi et al. 2008; Marcos et al. 2007). Enterocin 416K1 activated films reduced the levels of *L. monocytogenes* in contaminated frankfurters by ca. 1.5 to 0.5 log cycles within 24 h of storage at temperatures of 4 and 22 °C, but did not avoid exponential growth of the pathogen during further storage of samples (Iseppi et al. 2008). Marcos et al. (2007) tested the antilisterial effects of enterocins A and B immobilized in different supports (alignate, zein and polyvinyl alcohol) on air-packed and vacuum-packed sliced cooked ham stored at 6 °C. The most effective treatment for controlling *L. monocytogenes* during storage was vacuum-packaging of ham with alginate films containing 2,000 AU/cm² of enterocins, with no increase from inoculated levels of *L. monocytogenes* until day 15.

Pre-surface application of bacteriocins in combination with post-packaging treatments is another approach of recent interest. Bacteriocin application followed by in-package thermal treatments can provide an effective combination to control *L. monocytogenes* on products such as frankfurters or turkey bologna, as shown for pediocin, nisin, nisin-lysozyme, or combinations of these bacteriocins with sodium lactate/sodium diacetate (Chen et al. 2004a; Mangalassary et al. 2008). Mangalassary et al. (2008) studied the efficacy of in-package pasteurization (65 °C for 32 s.) combined with pre-surface application of nisin and/or lysozyme to reduce and prevent the subsequent recovery and growth of *L. monocytogenes* during refrigerated storage on the surface of low-fat turkey bologna. In-package pasteurization in combination with nisin or nisin–lysozyme treatments was effective in reducing the population below detectable levels by 2–3 weeks of storage. In bologna manufactured with different sodium lactate/sodium diacetate combinations, dipping in pediocin solution followed by heat treatment decreased the *D*-values for inactivation of *L. monocytogenes* at low pediocin concentration, but exhibited a protective effect at higher concentrations, indicating that interactions between additives in formulations can vary at different temperatures/concentrations (Maks et al. 2010). In a previous study, treatments of frankfurters with 3,000 AU or 6,000 AU pediocin (in ALTA 2341) followed by heating in hot water reduced the populations of inoculated *Listeria* in proportion to the intensity of treatments (Chen et al. 2004a). The combination of pediocin (6,000 AU) with post-packaging thermal treatment (81 °C or more for at least 60 s), achieved a 50 % reduction of initial inoculation levels. Little or no growth of *L. monocytogenes* was observed on the treated frankfurters for 12 weeks at 4 or 10 °C, and for 12 days at 25 °C. This treatment did not affect the sensory qualities of frankfurters. The authors of this study concluded that pediocin (in ALTA 2341) in combination with postpackaging thermal treatment offers an effective treatment combination for improved control of *L. monocytogenes* on frankfurters.

Another example of a combined treatment is the application of nisin with pulsed light. Application of a Nisaplin dip followed by exposure to pulsed light (PL; 9.4 J/cm²) reduced the population of *L. innocua* on sausages by 4 log cycles and inhibited its growth during refrigeration storage for 24–48 days (Uesugi and Moraru 2009). Since application of PL is approved for decontamination of food and food surfaces, the combined treatment could be applied as a post-processing step to reduce surface contamination and increase the safety of RTE meat products.

Bacteriocins have been proposed for use in packaged foods to increase the efficacy of irradiation treatments. One study reported that irradiation acted synergistically with pediocin on *L. monocytogenes* inoculated in packaged frankfurters (Chen et al. 2004b). Combination of pediocin with postpackaging irradiation at

1.2 kGy or more was necessary to achieve a 50 % reduction of *L. monocytogenes* on frankfurters. The combination of 6,000 AU of pediocin and irradiation at 2.3 kGy or more was the most effective treatment for inhibition of the pathogen for 12 weeks at 4 or 10 °C. Best results were reported on samples stored at 4 °C, with little or no growth of the pathogen during 12 weeks of storage and no adverse effects on the sensory quality of frankfurters. Similarly, bacteriocin-activated films have been tested as a way to increase the radiation sensitivity of the target pathogens, aimed at reducing radiation doses and impact on product quality. In ready-to-eat turkey meat vacuum-packaged with a pectin-nisin film and treated by low dose irradiation (2 kGy), the reduction obtained for the *L. monocytogenes* population (5.36 log CFU/cm^2) were greater compared to irradiation and pectin film single treatments. In addition, pectin-nisin films did significantly slow the proliferation of *L. monocytogenes* cells that survived irradiation during 8 weeks of storage at 10 °C (Jin et al. 2009). The authors concluded that the combined treatment could serve to prevent listeriosis due to postprocessing contamination while reducing radiation doses and impact on product quality, or to prevent *L. monocytogenes* growth in accidentally recontaminated packages of irradiated RTE meats.

High hydrostatic pressure processing (HHP) is now being used more frequently as a food processing technology that is applied on packaged foods. Several reports indicate that bacteriocins can enhance the antibacterial effects of HHP treatments. In one study, the efficacy of enterocins added to cooked ham increased in combination with a HHP treatment at 400 MPa for 10 min (Garriga et al. 2002). The combined treatment avoided overgrowth of *L. sakei* CTC746 strain during storage, improving the results compared to HHP treatment alone (Garriga et al. 2002). *L. monocytogenes* was also kept at levels <10 CFU/g for 61 days at 4 °C (Garriga et al. 2002). However, the bacteriocins had no effect on regrowth of other survivors (*Ln. carnosum* CTC747, *Staphylococcus carnosus* and *S. aureus* strains, *E. coli* or *S. enterica* strains). A combined treatment of enterocins (2,400 AU/g) and HHP (400 MPa, 10 min) avoided overgrowth of surviving listeria upon a simulated cold-chain break event when the samples were stored at 1 °C, but not at 6 °C (Marcos et al. 2008a), indicating the influence of storage temperature on the delicate balance between inhibited proliferation of survivors and repair of sublethal damage and cell growth.

Protective coatings in the form of activated films have also been tested to increase the efficacy of HHP in ready-to-eat meat products (Aymerich et al. 2008). The efficacy immobilized enterocins in combination with HHP to control *L. monocytogenes* growth during the shelf life of artificially inoculated cooked ham was investigated (Jofré et al. 2007). The antilisterial activity of enterocins immobilised in plastic interleaves was strongly potentiated by application of HHP treatment (400 MPa, 10 min), reducing viable counts by about 4 log units and holding the levels of *L. monocytogenes* in the treated sliced ham below 1.5 log CFU/g at the end of storage for 30 days at 6 °C (Jofré et al. 2007). Storage of samples at a lower temperature of 1 °C extended the protective effect of the combined treatment for at least 60 °C, even in the event of a simulated cold chain break (Marcos et al. 2008b). In a separate study (Marcos et al. 2008a), sliced cooked ham was packaged in alginate films containing or not enterocins A and B, and then was pressurized (400 MPa, 10 min, 17 °C).

While the single antimicrobial packaging treatment was able to inhibit growth of *L. monocytogenes* during the first 8 days of storage at 6 °C, and the single HHP pre-treatment attained a ca. 3.4 logs reduction of viable counts for about the same period followed by regrowth of the listeria in both cases, the combined treatment extended the lag phase of listeria to 22 days, and the slight growth observed afterwards did not exceed 1.8 log CFU/g by the end of storage (day 60). For samples stored at 1 °C, the combined treatment of HHP and enterocin film caused a faster decline of *L. monocytogenes* counts compared to HHP alone, but no regrowth was observed in either case for 60 days, suggesting that at the lower temperature of storage, antimicrobial packaging did not give additional protection against *L. monocytogenes* to pressurized samples. However, after a simulated cold-chain break event at day 60, there was a dramatic increase in the *L. monocytogenes* population for single HHP treatments (8.5 log CFU/g), indicating the capacity of pressure-injured *L. monocytogenes* cells to recover under favourable conditions. By contrast, for the combined treatment of HHP and enterocin films, the temperature abuse resulted in a slight increase until 1.7 log CFU/g at 90 days. The authors concluded the combination of antimicrobial packaging with HPP could be useful to control and reduce the numbers of *L. monocytogenes* and to overcome temperature abuse. In a similar study, Jofré et al. (2008) tested the effectiveness of the application of interleavers (composed by polypropylene/polyamide layers) containing enterocins A and B, sakacin K, nisin A, potassium lactate and nisin plus lactate alone or in combination with a 400 MPa HHP treatment in sliced cooked ham spiked with *Salmonella* spp. It was concluded that nisin was the only treatment that produced absence of *Salmonella* 24 h after pressurisation and the application of nisin through interleavers and combined with an HHP treatment appears as the most effective treatment to achieve absence of *Salmonella* in 25 g samples during refrigeration storage of the sliced ham (Jofré et al. 2008).

4.1.3   Fermented Meats

Bacteriocin preparations can be added to meat batters for reduction of the initial levels of bacteriocin-sensitive populations and inactivation of microbial pathogens in fermented meat products. The lower pH attained in sausages compared to fresh meats may increase the solubility of some bacteriocins like nisin, and probably their antimicrobial activity as well. Microbial inactivation by bacteriocin addition may also be an attractive hurdle for slightly fermented sausages, in which the higher pH and water content may facilitate survival and proliferation of certain pathogenic bacteria.

Several bacteriocins such as nisin, enterocins (CCM 4231, A, B and AS-48) or leucocins improved the reduction of *L. monocytogenes* or *S. aureus* populations in fermented meats (Rodríguez et al. 2002; Chen and Hoover 2003; Aymerich et al. 2008; Galvez et al. 2008). Addition of nisin alone was effective in preservation of bologna-type sausages against LAB spoilage (Davies and Delves-Broughton 1999) and in the inhibition of *L. monocytogenes* in sucuk, a Turkish fermented sausage

(Hampikyan and Ugur 2007). The effectiveness of nisin in fermented meats increased in combination with other antimicrobials, such as organic acids (reducing the viable counts of *S.* Kentucky and *S. aureus*; Scannell et al. 1997), lysozyme-EDTA (inhibiting the growth of *B. thermosphacta*, *L. curvatus*, *Ln. mesenteroides*, *L. monocytogenes* and *E. coli* O157:H7; Gill and Holley 2000a) or grape seed extract (Sivarooban et al. 2007). Enterocins can inhibit *Listeria* in fermented meats, as shown for enterocin CCM 4231 in dry fermented Hornád salami (Lauková et al. 1999) or enterocins A and B in espetec (traditional Spanish sausage; Aymerich et al. 2000). Addition of enterocin CCM 4231 (12,800 AU/g) from *E. faecium* CCM 4231 to Hornád salami meat mixture resulted in a reduction of *L. monocytogenes* by 1.67 log cycle immediately after addition of the bacteriocin (Lauková et al. 1999). Although the added bacteriocin did not prevent growth of the listeria during storage of samples in drying rooms at temperatures between 24 and 15 °C, viable counts were significantly lower that the controls. In espetec (a Spanish slightly-fermented sausage), addition of enterocins A and B (648 AU/g) reduced the viable counts of *L. innocua* below 50 CFU/g from the fifth day until the end of the process (12 days) of manufacturing (Aymerich et al. 2000).

In Italian sausages ("cacciatore"), enterocin 416K1 (10 AU/g, in the form of a concentrated culture supernatant) decreased the levels of *L. monocytogenes* in sausages by ca. 2.5 log CFU/g during the drying period (3 days), but failed to suppress the pathogen during ripening (Sabia et al. 2003). Regarding enterocin AS-48, after addition of this bacteriocin at 450 AU/g in a meat sausage model system, it was observed that no viable listeria were detected after 6 and 9 days of incubation at 20 °C (Ananou et al. 2005a), and also that viable counts of *S. aureus* were reduced below detectable levels at the end of storage (Ananou et al. 2005b). Also bacteriocins from leuconostocs have been tested in fermented meats. Addition of semi-purified bacteriocin of *Ln. mesenteroides* E131 improved the reduction of *L. monocytogenes* viable counts in challenge experiments during fermented sausage manufacturing (Drosinos et al. 2006).

4.2 Application of Protective Cultures

4.2.1 Raw Meats

Many LAB naturally associated with meats can grow at refrigeration temperatures. Therefore, bacteriocin-producing strains of these LAB that do not have adverse effects on meats can be selected as protective cultures for raw meat preservation (Table 4.2). Previous works have demonstrated the effectiveness of bacteriocin-producing *L. sakei* and *L. curvatus* strains in inhibiting *L. monocytogenes* or *B. thermosphacta* in raw meat products. When *L. sakei* CWBI-B1365 and *L. curvatus* CWBI-B28 (producers of sakacin G and P, respectively) were tested as protective cultures on raw beef and poultry meat challenged with *L. monocytogenes* and stored at 5 °C in sealed bags, inhibition of the listeria was found to depend greatly on the

Table 4.2 Examples of applications of bacteriocin-producing cultures in meat and poultry products

Starter or protective cultures		
Raw meats		
Bacteriocin producer *L. curvatus* CRL705	Effective inhibition of *L. innocua* and *B. thermosphacta* and indigenous contaminant LAB in fresh beef; contribution to meat ageing by limited proteolysis	Fadda et al. 2008
BLIS-producing *L. sakei*	Delayed blownpack spoilage caused by *C. estertheticum* and reduced survival of *C. jejuni* on meat	Jones et al. 2009
BLIS-producing *L. fermentum* ACA-DC179	Growth inhibition of *S. Enteritidis* in refrigerated chicken ground meat	Maragkoudakis et al. 2009
RTE meats		
Bacteriocin-producing *P. acidilactici* strains	Inhibition of *L. monocytogenes* in cooked meats	Rodríguez et al. 2002
Sakacin K-producing *L. sakei* CTC494	Inhibition of *L. monocytogenes* in cooked meat products	Hugas et al. 1998
Bacteriocin-producing *L. sakei*	Growth inhibition of *L. monocytogenes* and *E. coli* O157.H7 in cooked, sliced, vacuum-packaged meats	Bredholt et al. 1999
Fermented meats		
Bacteriocin-producing *L. sakei* starter cultures	Reduction of *Listeria* populations in fermented sausages	Ravyts et al. 2008
Curvacin-producing *L. curvatus*	Antilisterial effects in meat fermentation	Dicks et al. 2004
Pediocin-producing *P. acidilactici*	Commercial starter cultures for fermentation of meat products to reduce the numbers of *L. monocytogenes* in the final product	Amezquita and Brashears 2002
E. faecalis CECT7121 (producer of enterocin MR99)	Reduction of viable counts of *Enterobacteriaceae*, *S. aureus* and other Gram-positive cocci in craft dry-fermented sausages	Sparo et al. 2008

meat substrate (Dortu et al. 2008). On raw beef, *L. curvatus* CWBI-B28 was more effective in reducing *L. monocytogenes* cell concentrations below detectable levels (7 days) than *L. sakei* CWBI-B1365 (21 days). In poultry meat, the application of the LAB strains separately showed much lower inhibitory activities, but their addition in combination led to growth inhibition of the listeria. This is an interesting example of a synergistic effect between two sakacin-producing strains in a food system.

Lactobacillus curvatus CRL705 used as a protective culture in fresh beef was effective in inhibiting *L. innocua* and *B. thermosphacta* as well as the indigenous contaminant LAB at low temperatures and had a negligible effect on meat pH (Castellano et al. 2008). It was observed that meat inoculation with *L. curvatus* CRL705 showed a net increase of free amino acids, due to the complementary activity of the bacterial and meat proteases on meat sarcoplasmic proteins (Fadda et al. 2008).

It was proposed that *L. curvatus* CRL705 protective cultures could contribute to meat ageing by generating small peptides and free amino acids, while improving shelf life (Fadda et al. 2008).

Inoculation with a sakacin A producer *L. sakei* strain reduced the population of *L. monocytogenes* on vacuum-packed lamb during 12 week storage. Similarly, inoculation with BLIS-producing *L. sakei* strains delayed blownpack spoilage caused by *Clostridium estertheticum* and reduced the survival of *Campylobacter jejuni* on beef meat (Jones et al. 2009). In vacuum-packaged chicken cuts, inoculation with sakacin-P producing *L. sakei* achieved a growth inhibition of *L. monocytogenes* (Katla et al. 2002). Plantaricin-producing *L. plantarum* showed anti-listerial effects in uncooked and cooked chicken meat (Enan 2006; Gamal 2006). Enterococci have also been tested as protective cultures in raw meats. In chicken ground meat stored at 8–10 °C, growth of *L. monocytogenes* and *S.* Enteritidis was adversely affected by the respective presence of protective cultures consisting of strain *E. faecium* PCD71 (carrying the genetic determinants for enterocins A, P, L50A and L50B) and strain *L. fermentum* ACA-DC179, producer of BLIS against *Salmonella* (Maragkoudakis et al. 2009; Zoumpopoulou et al. 2008). Strain *E. faecium* PCD71 inhibited the growth of *L. monocytogenes* by at least 0. 7 log CFU/g after 7 days storage, while strain *L. fermentum* ACA-DC179 inhibited the growth of *S.* Enteritidis by up to 1.3 log CFU/g compared to the control (Maragkoudakis et al. 2009). In addition, none of these two strains caused detrimental effects on biochemical parameters related to spoilage of the chicken meat.

4.2.2 Semi-processed and Cooked Meats

Lactic acid bacteria are the prevalent spoilage microorganisms in cooked meat products (Mataragas et al. 2006, Audenaert et al. 2010, Chenoll et al. 2007). The shelf life of most heat processed meats is limited by *Lactobacillus* and *Leuconostoc* strains that rapidly recontaminate the product during handling and slicing (Lücke 2000). These LAB also tend to displace pathogenic bacteria. In the absence of competing microbiota, *L. monocytogenes* will proliferate more easily. Specific bacteriocin-producing LAB strains could be used as protective cultures for semi-processed and cooked meats provided that they cause only a minimal change in the desired sensory properties of the products while inhibiting *Listeria* and displacing other LAB involved in spoilage (Hugas et al. 1998; Lücke 2000; Chen and Hoover 2003; Aymerich et al. 2008; Galvez et al. 2008). Bacteriocin-producing protective cultures have been shown to inhibit *L. monocytogenes* in vacuum-packaged processed meats, such as *Lactobacillus bavaricus* MN in minimally heat-treated beef cubes (Winkowski et al. 1993), *P. acidilactici* JBL 1095 in wieners (Degnan et al. 1992), or *P. acidilactici* JD1-23 in frankfurters (Berry et al. 1991). In Brazillian raw sausage lingüiça, bacteriocin-producing *Lactobacillus sake* 2a also inhibited growth of *L. monocytogenes* (Liserre et al. 2002). The bacteriocinogenic strains *L. sakei* CTC494 and *E. faecium* CTC492 (producer of enterocins A and B) prevented slime formation in cooked pork by *Lb. sakei* but not by *Leuconostoc mesenteroides* (Aymerich et al. 1998). In sliced, vacuum-packaged cooked ham, the same enterococcal strain partially prevented ropiness by *L.sakei* (Aymerich et al. 2002). Inoculation of strains producing sakacin P or leucocin in cooked meat products was shown to inhibit growth of listeria (Katla et al. 2002; Jacobsen et al. 2003), and protective *L. sakei* cultures were also shown to inhibit *L. monocytogenes* and *E. coli* O157:H7 in vacuum-packed cooked meat products (Bredholt et al. 1999). The bacteriocinogenic strain *L. curvatus* CWBI-B28 reduced *L. monocytogenes* levels below detection limits in bacon meat within 1 or 2 weeks in absence or presence of nitrites, respectively (Ghalfi et al. 2006). Anti-listerial effect was also observed with a plantaricin producing *L. plantarum* strain in cooked chicken meat (Enan 2006). There are already several LAB cultures in the market introduced as starter or bioprotective culture with the aim of contributing to microbiological safety of semi-processed and cooked meats (Aymerich et al. 2008).

4.2.3 Fermented Meats

Certain lactic acid bacteria play key roles in meat fermentations. Therefore, bacteriocin-producing strains have been proposed as starter cultures to combat pathogens such as *L. monocytogenes* (Työppönen et al. 2003; Leroy et al. 2006; Aymerich et al. 2008). Bacteriocin-producing lactobacilli (mainly *L. sakei* and *L. curvatus*, but also *Lactobacillus rhamnosus* and *L. plantarum*) have demonstrated anti-listerial effects in sausage or salami fermentations, depending to a great extent on strain and type of meat (Erkkilä et al. 2001; Leroy et al. 2005; Dicks et al. 2004; Benkerroum et al. 2005; Todorov et al. 2007) (Table 4.2).

L. sakei CTC 494 (producing sakacin K) is a promising functional starter culture with antilisterial activity, being capable to successfully suppress *L. monocytogenes* in Spanish-style and German-style fermented sausages (Aymerich et al. 2008) or to reduce listeria populations in Belgian-style sausages, Italian salami, and Cacciatore salami (Ravyts et al. 2008). The efficacy of *L. sakei* is influenced by environmental factors such as sausage ingredients, salt, fat and nitrite content, acidification level, and temperature (Leroy et al. 2006). Since *L. sakei* and *L. curvatus* can hydrolyze muscle sarcoplasmic proteins and, in a lesser extent, myofibrillar proteins, they can contribute to the generation of small peptides and amino acids which contribute as direct flavour enhancers or as precursors of other flavour compounds during the ripening of dry-fermented sausages (Leroy et al. 2006). Exploitation of these activities may lead to the use of a new generation starter cultures with industrial or nutritional important functionalities (Leroy et al. 2006). Another, yet unexplored possible application of these functional properties would be the generation of bioactive peptides from the meat proteins by selected LAB with adequare proteolytic activities.

Bacteriocin-producing pediococci can reduce *L. monocytogenes* populations in fermented meats (Amezquita and Brashears 2002; Rodríguez et al. 2002; Aymerich et al. 2008). Pediococci are preferred as starters in certain products (rather than lactobacilli), e.g. in American-style sausages fermented at higher temperatures. Bacteriocin-producing pediococci were proposed as indigenous starter cultures in the fermentation of Urutan, a Balinese traditional dry fermented sausage (Antara et al. 2004). One advantage is that pediocin PA-1 producers do not inhibit bacteria relevant to the fermentation such as staphylococci and micrococci (Gonzalez and Kunka 1987).

Enterococci are often part of the normal microbiota in meat fermentations, and have demonstrated to be effective as antilisteria agents in fermented meats, being also able to inhibit *S. aureus* (Foulquié Moreno et al. 2003; Aymerich et al. 2008; Galvez et al. 2008). However, their application in foods is controversial because of their potential virulence as opportunistic pathogens and also as carriers of antimicrobial resistance genes. The bacteriocinogenic strains *E. faecium* CCM 4231 and *E. faecium* RZS C13 strongly inhibited the growth of *Listeria* spp. in sausage fermentations (Callewaert et al. 2000), and *Enterococcus casseliflavus* IM 416K1 (producer of enterocin 416K1) was able to suppress *L. monocytogenes* in artificially inoculated "cacciatore" Italian sausages (Sabia et al. 2003). During sausage fermentation, inoculated *Enterococcus faecalis* A-48-32 (producer of the broad-spectrum cyclic enterocin AS-48) or its transconjungant *E. faecium* S-32-81, reduced the concentration of *L. monocytogenes* down to undetactable levels within 7 or 6 days of incubation at 20 °C (Ananou et al. 2005a). Similarly, strain A-48-32 inhibited growth of *S. aureus* and reduced viable cell counts to 1 log CFU/g at the end of fermentation (Ananou et al. 2005b). Strain *E. faecalis* CECT 7121 (isolated from natural corn silage, and producer of the broad-spectrum enterocin MR99) is interesting because it is devoid of the genes for haemolysin and gelatinase production, and does not produce biogenic amines (Sparo et al. 2008). When tested in the manufacture of craft dry-fermented sausages, the sausages inoculated with *E. faecalis* CECT 7121 had lower viable counts of *Enterobacteriaceae*, *S. aureus* and other Gram-positive cocci at the end of fermentation (2 days), with no detectable enterobacteria and *S. aureus* at the end of drying (21 days). *E. faecalis* CECT7121 did not affect the growth of *Lactobacillus* spp. but it displaced the autochthonous populations of enterococci (Sparo et al. 2008).

The potential of bacteriocin-producing lactococci in meat fermentations has been studied to a much less extent. Nisin-producing lactococcal strains isolated from fermented sausages were suggested as adjunct cultures for improving the food safety of meat fermented products manufactured under poor hygienic conditions such as indigenous fermentations (Rodriguez et al. 1995; Noonpakdee et al. 2003). Furthermore, it was reported that a transformant *L. lactis* strain producing lacticin 3417 significantly reduced the populations of *L. innocua* and *S. aureus* in sausages, although growth of the bacteriocin producer was markedly influenced by sausage ingredients (Scannell et al. 2001). In another study on manufacture of merguez, a dry-fermented beef meat sausage, inoculation with the Bac + strain *L. lactis* subsp. *lactis* M significantly reduced the levels of *L. monocytogenes* during the fermentation phase (Benkerroum et al. 2003). However, inoculation with a lyophilized culture of the bacteriocin-producing strain *L. lactis* LMG21206 decreased *Listeria* counts to below the detectable limit after 15 days of drying, but it had no effect on the viability of the listeria during sausage fermentation. By comparison, the results obtained with the Bac + strain *L. curvatus* LBPE were superior, with highly significant reductions during fermentation and ripening (Benkerroum et al. 2005).

Several LAB strains may antagonise growth o *E. coli* O157:H7 in fermented sausages. This inhibitory effect has been attributed to the production of small antimicrobial compounds (such as reuterin, 3-hydroxy fatty acids, phenyllactic acid, and 4-hydroxyphenyllactic acid and novel bacteriocins; Leroy et al. 2006). It was shown that inoculation of salami with strains of *Lactobacillus* spp. as well as bifidobacteria reduced the levels of *L. monocytogenes* and *E. coli* O111 during fermentation of sausage batter (Pidcock et al. 2002). Similar results were reported for *Lactobacillus reuteri* and *Bifidobacterium longum* in dry fermented sausages. In the treatment containing *L. reuteri* (producer of reuterin), a 3 log CFU/g reduction in *E. coli* O157:H7 numbers was found at the end of drying, while *B. longun* was reported to have lower effects (1.9 log CFU/g reduction) (Muthukumarasamy and Holley 2007).

Staphylococci and micrococci may also be exploited as sources for antibacterial substances applicable in sausage fermentations. The introduction of the lysostaphin gene (an endopeptidase that specifically cleaves the glycine–glycine bonds unique to the interpeptide cross-bridge of the *S. aureus* cell wall) into meat starter lactobacilli (Cavadini et al. 1998) is an interesting approach to prevent the growth of *S. aureus*. Furthermore, one *Staphylococcus xylosus* sausage isolate that produces an antilisterial substance increased the microbial inactivation of *L. monocytogenes* in Naples-type sausage (Villani et al. 1997). *Kocuria varians* (formerly *Micrococcus varians*) produces the lantibiotic variacin (Pridmore et al. 1996). Strains producing this lantibiotic were isolated form Italian-type raw salami fermentations. Bacteriocinogenic *Kocuria* strains could be very interesting as adjunct protective cultures in meat fermentations.

References

Amezquita A, Brashears MM (2002) Competitive inhibition of *Listeria monocytogenes* in ready-to-eat meat products by lactic acid bacteria. J Food Prot 65:316–325

Ananou S, Garriga M, Hugas M et al (2005a) Control of *Listeria monocytogenes* in model sausages by enterocin AS-48. Int J Food Microbiol 103:179–190

Ananou S, Maqueda M, Martínez-Bueno M et al (2005b) Control of *Staphylococcus aureus* in sausages by enterocin AS-48. Meat Sci 71:549–576

Antara NS, Sujaya IN, Yokota A et al (2004) Effects of indigenous starter cultures on the microbial and physicochemical characteristics of Urutan, a Balinese fermented sausage. J Biosci Bioeng 98:92–98

Ariyapitipun T, Mustapha A, Clarke AD (1999) Microbial shelf life determination of vacuum-packaged fresh beef treated with polylactic acid, lactic acid, and nisin solutions. J Food Prot 62:913–920

Ariyapitipun T, Mustapha A, Clarke AD (2000) Survival of *Listeria monocytogenes* Scott A on vacuum-packaged raw beef treated with polylactic acid, lactic acid, and nisin. J Food Prot 63:131–136

Audenaert K, D'Haene K, Messens K et al (2010) Diversity of lactic acid bacteria from modified atmosphere packaged sliced cooked meat products at sell-by date assessed by PCR-denaturing gradient gel electrophoresis. Food Microbiol 27:12–18

Aymerich MT, Garriga M, Costa S et al (2002) Prevention of ropiness in cooked pork by bacteriocinogenic cultures. Int Dairy J 12:239–246

Aymerich MT, Hugas M, Monfort JM (1998) Review: bacteriocinogenic lactic acid bacteria associated with meat products. Food Sci Technol Int 4:141–158

Aymerich T, Garriga M, Ylla J (2000) Application of enterocins as biopreservatives against *Listeria innocua* in meat products. J Food Prot 63:721–726

Aymerich T, Picouet PA, Monfort JM (2008) Decontamination technologies for meat products. Meat Sci 78:114–129

Azuma T, Bagenda DK, Yamamoto T et al (2007) Inhibition of *Listeria monocytogenes* by freeze-dried piscicocin CS526 fermentate in food. Lett Appl Microbiol 44:138–144

Barboza de Martinez Y, Ferrer K, Salas EM (2002) Combined effects of lactic acid and nisin solution in reducing levels of microbiological contamination in red meat carcasses. J Food Prot 65:1780–1783

Benkerroum N, Baoudi A, Kamal M (2003) Behaviour of *Listeria monocytogenes* in raw sausages (merguez) in presence of a bacteriocin producing lactococcal strain as a protective culture. Meat Sci 63:479–484

Benkerroum N, Daoudi A, Hamraoui T et al (2005) Lyophilized preparations of bacteriocinogenic *Lactobacillus curvatus* and *Lactococcus lactis* subsp. *lactis* as potential protective adjuncts to control *Listeria monocytogenes* in dry-fermented sausages. J Appl Microbiol 98:56–63

Berry ED, Hutkins RW, Mandigo RW (1991) The use of bacteriocin-producing *Pediococcus acidilactici* to control postprocessing *Listeria monocytogenes* contamination of frankfurters. J Food Prot 54:681–686

Björkroth J, Korkeala H (1997) Ropy slime-producing *Lactobacillus sake* strains possess a strong competitive ability against a commercial biopreservative. Int J Food Microbiol 38:117–123

Blanco M, Massani MR, Fernandez A et al (2008) Development and characterization of an active polyethylene film containing *Lactobacillus curvatus* CRL705 bacteriocins. Food Addit Contam 25:1424–1430

Blanco M, Massani P, Morando G et al (2012) Characterization of a multilayer film activated with *Lactobacillus curvatus* CRL705 bacteriocins. J Sci Food Agric 92:1318–1323

Blanco Massani M, Molina V, Sanchez M et al (2014) Active polymers containing *Lactobacillus curvatus* CRL705 bacteriocins: effectiveness assessment in Wieners. Int J Food Microbiol 178:7–12

Borch E, Kant-Muermans ML, Blixt Y (1996) Bacterial spoilage of meat and cured meat product. Int J Food Microbiol 33:103–120

Bredholt S, Nesbakken T, Holck A (1999) Protective cultures inhibit growth of *Listeria monocytogenes* and *Escherichia coli* O157:H7 in cooked, sliced, vacuum- and gas-packaged meat. Int J Food Microbiol 53:43–52

Callewaert R, Hugas M, De Vuyst L (2000) Competitiveness and bacteriocin production of enterococci in the production of Spanish-style dry fermented sausages. Int J Food Microbiol 57:33–42

Castellano P, Belfiore C, Fadda S et al (2008) A review of bacteriocinogenic lactic acid bacteria used as bioprotective cultures in fresh meat produced in Argentina. Meat Sci 79:483–499

Castellano P, Vignolo G (2006) Inhibition of *Listeria innocua* and *Brochothrix thermosphacta* in vacuum-packaged meat by addition of bacteriocinogenic Lactobacillus curvatus CRL705 and its bacteriocins. Lett Appl Microbiol 43:194–199

Cavadini C, Hertel C, Hammes WP (1998) Application of lysostaphin-producing lactobacilli to control staphylococcal food poisoning in meat products. J Food Prot 61:419–424

Chen CM, Sebranek JG, Dickson JS et al (2004a) Combining pediocin (ALTA 2341) with post-packaging thermal pasteurization for control of *Listeria monocytogenes* on frankfurters. J Food Prot 67:1855–1865

Chen CM, Sebranek JG, Dickson JS et al (2004b) Combining pediocin with postpackaging irradiation for control of *Listeria monocytogenes* on frankfurters. J Food Prot 67:1866–1875

Chen H, Hoover DG (2003) Bacteriocins and their food applications. Comp Rev Food Sci Food Safety 2:82–100

Chenoll E, Macián MC, Elizaquível P et al (2007) Lactic acid bacteria associated with vacuum-packed cooked meat product spoilage: population analysis by rDNA-based methods. J Appl Microbiol 102:498–508

Coma V (2008) Bioactive packaging technologies for extended shelf life of meat-based products. Meat Sci 78:90–103

Cosby DE, Harrison MA, Toledo RT (1999) Vacuum or modified atmosphere packaging and EDTA-nisin treatment to increase poultry product shelf-life. J Appl Poultry Res 8:185–190

Cutter CN, Siragusa GR (1996a) Reductions of *Listeria innocua* and *Brochothrix thermosphacta* on beef following nisin spray treatments and vacuum packaging. Food Microbiol 13:23–33

Cutter CN, Siragusa GR (1996b) Reduction of *Brochothrix thermosphacta* on beef surfaces following immobilization of nisin in calcium alginate gels. Lett Appl Microbiol 23:9–12

Davies EA, Delves-Broughton J (1999) Nisin. In: Robinson R, Batt C, Patel P (eds) Encyclopedia of food microbiology. Academic Press, London, pp 191–198

Degnan A, Luchansky J (1992) Influence of beef tallow and muscle on the antilisterial activity of pediocin AcH and liposome-encapsulated pediocin AcH. J Food Prot 55:552–554

Degnan AJ, Byong N, Luchansky JB (1993) Antilisterial activity of pediocin AcH in model food systems in the presence of an emulsifier or encapsulated within liposomes. Int J Food Microbiol 18:127–138

Degnan AJ, Yousef AE, Luchansky JB (1992) Use of *Pediococcus acidilactici* to control *Listeria monocytogenes* in temperature abused vacuum-packaged wieners. J Food Prot 55:98–103

Dicks LMT, Mellett FD, Hoffman LC (2004) Use of bacteriocin-producing starter cultures of *Lactobacillus plantarum* and *Lactobacillus curvatus* in production of ostrich meat salami. Meat Sci 66:703–708

Dortu C, Huch M, Holzapfel WH et al (2008) Anti-listerial activity of bacteriocin-producing *Lactobacillus curvatus* CWBI-B28 and *Lactobacillus sakei* CWBI-B1365 on raw beef and poultry meat. Lett Appl Microbiol 47:581–586

Drosinos EH, Mataragas M, Veskovic-Moracanin S et al (2006) Quantifying nonthermal inactivation of *Listeria monocytogenes* in European fermented sausages using bacteriocinogenic lactic acid bacteria or their bacteriocins: a case study for risk assessment. J Food Prot 69:2648–2663

Economou T, Pournis N, Ntzimani A et al (2009) Nisin-EDTA treatments and modified atmosphere packaging to increase fresh chicken meat shelf-life. Food Chemist 114:1470–1476

Enan G (2006) Behaviour of *Listeria monocytogenes* LMG 10470 in poultry meat and its control by the bacteriocin plantaricin UG 1. Int J Poultry Sci 5:355–359

Enan G, Alalyan S, Abdel-salam HA et al (2002) Inhibition of *Listeria monocytogenes* LMG10470 by plantaricin UG1 in vitro and in beef meat. Nahrung 46:411–414

Ercolini D, Ferrocino I, La Storia A et al (2010) Development of spoilage microbiota in beef stored in nisin activated packaging. Food Microbiol 27:137–143

Ercolini D, Storia A, Villani F et al (2006) Effect of a bacteriocin-activated polythene film on *Listeria monocytogenes* as evaluated by viable staining and epifluorescence microscopy. J Appl Microbiol 100:765–772

Erkkilä S, Suihko ML, Eerola S et al (2001) Dry sausage fermented by *Lactobacillus rhamnosus* strains. Int J Food Microbiol 64:205–210

Fadda S, Chambon C, Champomier-Vergès MC et al (2008) *Lactobacillus* role during conditioning of refrigerated and vacuum-packaged Argentinean meat. Meat Sci 79:603–610

Fang TJ, Lin LW (1994a) Growth of *Listeria monocytogenes* and *Pseudomonas fragi* on cooked pork in a modified atmosphere packaging/nisin combination system. J Food Prot 57:479–485

Fang TJ, Lin LW (1994b) Inactivation of *Listeria monocytogenes* on raw pork treated with modified atmosphere packaging and nisin. J Food Drug Anal 2:189–200

Ferrocino I, La Storia A, Torrieri E et al (2013) Antimicrobial packaging to retard the growth of spoilage bacteria and to reduce the release of volatile metabolites in meat stored under vacuum at 1 °C. J Food Prot 76:52–58

Foulquié Moreno MR, Rea MC, Cogan TM et al (2003) Applicability of a bacteriocin-producing *Enterococcus faecium* as a co-culture in Cheddar cheese manufacture. Int J Food Microbiol 81:73–84

Franklin NB, Cooksey KD, Getty KJ (2004) Inhibition of *Listeria monocytogenes* on the surface of individually packaged hot dogs with a packaging film coating containing nisin. J Food Prot 67:480–485

Gálvez A, Abriouel H, López RL et al (2007) Bacteriocin-based strategies for food biopreservation. Int J Food Microbiol 120:51–70

Galvez A, Lucas R, Abriouel H et al (2008) Application of bacteriocins in the control of foodborne pathogenic and spoilage bacteria. Crit Rev Biotechnol 28:125–152

Gamal E (2006) Behaviour of *Listeria monocytogenes* LMG 10470 in poultry meat and its control by the bacteriocin plantaricin UG 1. Int J Poultry Sci 5:355–359

Garriga M, Aymerich MT, Costa S et al (2002) Bactericidal synergism through bacteriocins and high pressure in a meat model system during storage. Food Microbiol 19:509–518

Ghalfi H, Kouakou P, Duroy M et al (2006) Antilisterial bacteriocin-producing strain of *Lactobacillus curvatus* CWBI-B28 as a preservative culture in bacon meat and influence of fat and nitrites on bacteriocins production and activity. Food Sci Technol Int 12:325–333

Gill AO, Holley RA (2000a) Inhibition of bacterial growth on ham and bologna by lysozyme, nisin and EDTA. Food Res Int 33:83–90

Gill AO, Holley RA (2000b) Surface application of lysozyme, nisin, and EDTA to inhibit spoilage and pathogenic bacteria on ham and bologna. J Food Prot 63:1338–1346

Goff JH, Bhunia AK, Johnson MG (1996) Complete inhibition of low levels of *Listeria monocytogenes* on refrigerated chicken meat with pediocin AcH bound to heat-killed *Pediococcus acidilactici* cells. J Food Prot 59:1187–1192

Gonzalez CF, Kunka BS (1987) Plasmid-associated bacteriocin production and sucrose fermentation in *Pediococcus acidilactici*. Appl Environ Microbiol 53:2534–2538

Govaris A, Solomakos N, Pexara A et al (2010) The antimicrobial effect of oregano essential oil, nisin and their combination against *Salmonella enteritidis* in minced sheep meat during refrigerated storage. Int J Food Microbiol 137:175–180

Hampikyan H, Ugur M (2007) The effect of nisin on *L. monocytogenes* in Turkish fermented sausages (sucuks). Meat Sci 76:327–332

Hugas M, Pagés F, Garriga M et al (1998) Application of the bacteriocinogenic *Lactobacillus sakei* CTC494 to prevent growth of *Listeria* in fresh and cooked meat products packaged with different atmospheres. Food Microbiol 15:639–650

Iseppi R, Pilati F, Marini M et al (2008) Anti-listerial activity of a polymeric film coated with hybrid coatings doped with Enterocin 416K1 for use as bioactive food packaging. Int J Food Microbiol 123:281–287

Jackson TC, Acuff GR, Dickson JS (1997) Meat, poultry and seafood. In: Doyle MP, Beuchat TJ (eds) Food microbiology: fundamentals and frontiers. ASM Press, Washington, DC, pp 83–100

Jacobsen T, Budde BB, Koch AG (2003) Application of *Leuconostoc carnosum* for biopreservation of cooked meat products. J Appl Microbiol 95:242–249

Jin T, Liu L, Sommers CH et al (2009) Radiation sensitization and postirradiation proliferation of *Listeria monocytogenes* on ready-to-eat deli meat in the presence of pectin-nisin films. J Food Prot 72:644–649

Jofré A, Aymerich T, Garriga M et al (2008) Assessment of the effectiveness of antimicrobial packaging combined with high pressure to control *Salmonella* sp. in cooked ham. Food Control 19:634–638

Jofré A, Garriga M, Aymerich T (2007) Inhibition of *Listeria monocytogenes* in cooked ham through active packaging with natural antimicrobials and high-pressure processing. J Food Prot 70:2498–2502

Jones RJ, Zagorec M, Brightwell G et al (2009) Inhibition by *Lactobacillus sakei* of other species in the flora of vacuum packaged raw meats during prolonged storage. Food Microbiol 26:876–881

Kalchayanand N (1990) Extension of shelf-life of vacuum-packaged refrigerated fresh beef by bacteriocins of lactic acid bacteria. Ph.D. Thesis, University of Wyoming

Katla T, Møretrø T, Sveen I et al (2002) Inhibition of *Listeria monocytogenes* in chicken cold cuts by addition of sakacin P and sakacin P-producing *Lactobacillus sakei*. J Appl Microbiol 93:191–196

Korkeala H, Suortti T, Mäkelä P (1988) Ropy slime formation in vacuum-packed cooked meat products caused by homofermentative lactobacilli and a *Leuconostoc* species. Int J Food Microbiol 7:339–347

Labadie J (1999) Consequences of packaging on bacterial growth. Meat is an ecological niche. Meat Sci 52:299–305

Lauková A, Czikková S, Laczková S et al (1999) Use of enterocin CCM 4231 to control *Listeria monocytogenes* in experimentally contaminated dry fermented Hornád salami. Int J Food Microbiol 52:115–119

Leroy F, Lievens K, De Vuyst L (2005) Interactions of meat-associated bacteriocin-producing Lactobacilli with *Listeria innocua* under stringent sausage fermentation conditions. J Food Prot 68:2078–2084

Leroy F, Verluyten J, De Vuyst L (2006) Functional meat starter cultures for improved sausage fermentation. Int J Food Microbiol 106:270–285

Liserre AM, Landgraf M, Destro MT et al (2002) Inhibition of *Listeria monocytogenes* by a bacteriocinogenic *Lactobacillus sake* strain in modified atmosphere-packaged Brazilian sausage. Meat Sci 61:449–455

Luchansky JB, Call JE (2004) Evaluation of nisin-coated cellulose casings for the control of *Listeria monocytogenes* inoculated onto the surface of commercially prepared frankfurters. J Food Prot 67:1017–1021

Lücke FK (2000) Utilization of microbes to process and preserve meat. Meat Sci 56:105–115

Lungu B, Johnson MG (2005) Fate of *Listeria monocytogenes* inoculated onto the surface of model Turkey frankfurter pieces treated with zein coatings containing nisin, sodium diacetate, and sodium lactate at 4 °C. J Food Prot 68:855–859

Maks N, Zhu L, Juneja VK et al (2010) Sodium lactate, sodium diacetate and pediocin: effects and interactions on the thermal inactivation of *Listeria monocytogenes* on bologna. Food Microbiol 27:64–69

Mangalassary S, Han I, Rieck J et al (2008) Effect of combining nisin and/or lysozyme with in-package pasteurization for control of *Listeria monocytogenes* in ready-to-eat turkey bologna during refrigerated storage. Food Microbiol 25:866–870

Maragkoudakis PA, Mountzouris KC, Psyrras D et al (2009) Functional properties of novel protective lactic acid bacteria and application in raw chicken meat against *Listeria monocytogenes* and *Salmonella enteritidis*. Int J Food Microbiol 130:219–226

Marcos B, Aymerich T, Monfort JM et al (2007) Use of antimicrobial biodegradable packaging to control *Listeria monocytogenes* during storage of cooked ham. Int J Food Microbiol 120:152–158

Marcos B, Aymerich T, Monfort JM et al (2008a) High-pressure processing and antimicrobial biodegradable packaging to control *Listeria monocytogenes* during storage of cooked ham. Food Microbiol 25:177–182

Marcos B, Jofré A, Aymerich T et al (2008b) Combined effect of natural antimicrobials and high pressure processing to prevent *Listeria monocytogenes* growth after a cold chain break during storage of cooked ham. Food Control 19:76–81

Mataragas M, Drosinos EH, Vaidanis A et al (2006) Development of a predictive model for spoilage of cooked cured meat products and its validation under constant and dynamic temperature storage conditions. J Food Sci 71:157–167

Mattila K, Saris P, Työppönen S (2003) Survival of *Listeria monocytogenes* on sliced cooked sausage after treatment with pediocin AcH. Int J Food Microbiol 89:281–286

Mauriello G, Ercolini D, La Storia A et al (2004) Development of polythene films for food packaging activated with an antilisterial bacteriocin from *Lactobacillus curvatus* 32Y. J Appl Microbiol 97:314–322

McMillin KW (2008) Where is MAP Going? A review and future potential of modified atmosphere packaging for meat. Meat Sci 80:43–65

Millette M, Le Tien C, Smoragiewicz W et al (2007) Inhibition of *Staphylococcus aureus* on beef by nisin-containing modified alginate films and beads. Food Control 18:878–884

Ming X, Weber GH, Ayres JM et al (1997) Bacteriocins applied to food packaging materials to inhibit *Listeria monocytogenes* on meats. J Food Sci 62:413–414

Moon SH, Paik HD, White S et al (2011) Influence of nisin and selected meat additives on the antimicrobial effect of ovotransferrin against *Listeria monocytogenes*. Poult Sci 90:2584–2591

Motlagh AM, Holla S, Johnson MC et al (1992) Inhibition of *Listeria* spp. in sterile food systems by pediocin AcH, a bacteriocin produced by *Pediococcus acidilactici* H. J Food Prot 55:337–343

Murphy RY, Hanson RE, Feze N et al (2005) Eradicating *Listeria monocytogenes* from fully cooked franks by using an integrated pasteurization-packaging system. J Food Prot 68:507–511

Murray M, Richard JA (1997) Comparative study of the antilisterial activity of Nisin A and Pediocin AcH in fresh ground pork stored aerobically at 5°C. J Food Prot 60:1534–1540

Muthukumarasamy P, Holley RA (2007) Survival of *Escherichia coli* O157:H7 in dry fermented sausages containing micro-encapsulated probiotic lactic acid bacteria. Food Microbiol 24:82–88

Natrajan N, Sheldon BW (1995) Evaluation of bacteriocin-based packaging and edible film delivery sistem to reduce *Salmonella* in fresh poultry. Poultry Sci 74:31

Natrajan N, Sheldon BW (2000a) Efficacy of nisin-coated polymer films to inactivate *Salmonella* Typhimurium on fresh broiler skin. J Food Prot 63:1189–1196

Natrajan N, Sheldon BW (2000b) Inhibition of *Salmonella* on poultry skin using protein- and polysaccharide-based films containing a nisin formulation. J Food Prot 63:1268–1272

Nattress FM, Baker LP (2003) Effects of treatment with lysozyme and nisin on the microflora and sensory properties of commercial pork. Int J Food Microbiol 85:259–267

Nattress FM, Yost CK, Baker LP (2001) Evaluation of the ability of lysozyme and nisin to control meat spoilage bacteria. Int J Food Microbiol 70:111–119

Nguyen VT, Gidley MJ, Dykes GA (2008) Potential of a nisin-containing bacterial cellulose film to inhibit *Listeria monocytogenes* on processed meats. Food Microbiol 25:471–478

Nielsen JW, Dickson JS, Crouse JD (1990) Use of a bacteriocin produced by *Pediococcus acidilactici* to inhibit *Listeria monocytogenes* associated with fresh meat. Appl Environ Microbiol 56:2142–2145

Nieto-Lozano JC, Reguera-Useros JI, Peláez-Martínez MC et al (2006) Effect of a bacteriocin produced by *Pediococcus acidilactici* against *Listeria monocytogenes* and *Clostridium perfringens* on Spanish raw meat. Meat Sci 72:57–61

Noonpakdee W, Santivarangkna C, Jumriangrit P et al (2003) Isolation of nisin-producing *Lactococcus lactis* WNC 20 strain from ham, a traditional Thai fermented sausage. Int J Food Microbiol 81:137–145

Nychas GJE, Skandamis PN, Tassou CC et al (2008) Meat spoilage during distribution. Meat Sci 78:77–89

O'Sullivan L, Ross RP, Hill C (2002) Potential of bacteriocin-producing lactic acid bacteria for improvements in food safety and quality. Biochimie 84:593–604

Pawar DD, Malik SVS, Bhilegaonkar KN et al (2000) Effect of nisin and its combination with sodium chloride on the survival of *Listeria monocytogenes* added to raw buffalo meat mince. Meat Sci 56:215–219

Pidcock K, Heard GM, Henriksson A (2002) Application of nontraditional meat starter cultures in production of Hungarian salami. Rev Int J Food Microbiol 76:75–81

Pridmore D, Rekhif N, Pittet AC et al (1996) Variacin, a new lanthionine-containing bacteriocin produced by *Micrococcus varians*: comparison to lacticin 481 of *Lactococcus lactis*. Appl Environ Microbiol 62:1799–1802

Quintavalla S, Vicini L (2002) Antimicrobial food packaging in meat industry. Meat Sci 62:373–380

Ravyts F, Barbuti S, Frustoli MA et al (2008) Competitiveness and antibacterial potential of bacteriocin-producing starter cultures in different types of fermented sausages. J Food Prot 71:1817–1827

Rodriguez JM, Cintas LM, Casaus P et al (1995) Isolation of nisin-producing *Lactococcus lactis* strains from dry fermented sausages. J Appl Bacteriol 78:109–115

Rodríguez JM, Martinez MI, Kok J (2002) Pediocin PA-1, a wide-spectrum bacteriocin from lactic acid bacteria. Crit Rev Food Sci Nutr 42:91–121

Sabia C, de Niederhäusern S, Messi P et al (2003) Bacteriocin-producing *Enterococcus casseliflavus* IM 416K1, a natural antagonist for control of *Listeria monocytogenes* in Italian sausages ("cacciatore"). Int J Food Microbiol 87:173–179

Santiago-Silva P, Nilda FF, Soares Juliana E et al (2009) Antimicrobial efficiency of film incorporated with pediocin (ALTA® 2351) on preservation of sliced ham. Food Control 20:85–89

Scannell AG, Ross RP, Hill C et al (2000a) An effective lacticin biopreservative in fresh pork sausage. J Food Prot 63:370–375

Scannell AGM, Hill C, Buckley DJ et al (1997) Determination of the influence of organic acids and nisin on shelf-life and microbiological safety aspects of fresh pork sausage. J Appl Microbiol 83:407–412

Scannell AGM, Hill C, Ross RP et al (2000b) Development of bioactive food packaging materials using immobilised bacteriocins Lacticin 3147 and Nisaplin®. Int J Food Microbiol 60:241–249

Scannell AGM, Schwarz G, Hill C et al (2001) Preinoculation enrichment procedure enhances the performance of bacteriocinogenic *Lactococcus lactis* meat starter culture. Int J Food Microbiol 64:151–159

Schlyter JH, Glass KA, Loeffelholz J et al (1993) The effects of diacetate with nitrite, lactate, or pediocin on the viability of *Listeria monocytogenes* in turkey slurries. Int J Food Microbiol 19:271–281

Siragusa GR, Cutter CN, Willett JL (1999) Incorporation of bacteriocin in plastic retains activity and inhibits surface growth of bacteria on meat. Food Microbiol 16:229–235

Sivarooban T, Hettiarachchy NS, Johnson MG (2007) Inhibition of *Listeria monocytogenes* using nisin with grape seed extract on turkey frankfurters stored at 4 and 10 °C. J Food Prot 70:1017–1020

Solomakos N, Govaris A, Koidis P et al (2008) The antimicrobial effect of thyme essential oil, nisin, and their combination against *Listeria monocytogenes* in minced beef during refrigerated storage. Food Microbiol 25:120–127

Sparo M, Nuñez GG, Castro M et al (2008) Characteristics of an environmental strain, *Enterococcus faecalis* CECT7121, and its effects as additive on craft dry-fermented sausages. Food Microbiol 25:607–615

Stergiou VA, Thomas LV, Adams MR (2006) Interactions of nisin with glutathione in a model protein system and meat. J Food Prot 69:951–956

Taalat E, Yousef AE, Ockerman HW (1993) Inactivation and attachment of *Listeria monocytogenes* on beef muscle treated with lactic acid and selected bacteriocins. J Food Prot 56:29–33

Thomas LV, Clarkson MR, Delves-Broughton J (2000) Nisin. In: Naidu AS (ed) Natural food antimicrobial systems. CRC-Press, Boca Raton, FL, pp 463–524

Todorov SD, Koep KSC, Van Reenen CA et al (2007) Production of salami from beef, horse, mutton, Blesbok (Damaliscus dorcas phillipsi) and Springbok (Antidorcas marsupialis) with bacteriocinogenic strains of *Lactobacillus plantarum* and *Lactobacillus curvatus*. Meat Sci 77:405–412

Työppönen S, Petäjä E, Mattila-Sandholm T (2003) Bioprotectives and probiotics for dry sausages. Int J Food Microbiol 83:233–244

Uesugi AR, Moraru CI (2009) Reduction of *Listeria* on ready-to-eat sausages after exposure to a combination of pulsed light and nisin. J Food Prot 72:347–353

Uhart ML, Ravishankar S, Maks ND (2004) Control of *Listeria monocytogenes* with combined antimicrobials on beef franks stored at 4 °C. J Food Prot 67:2296–2301

Villani F, Sannino L, Moschetti G et al (1997) Partial characterization of an antagonistic substance produced by *Staphylococcus xylosus* 1E and determination of the effectiveness of the producer strain to inhibit *Listeria monocytogenes* in Italian sausages. Food Microbiol 14:555–566

Wijnker JJ, Weerts EAWS, Breukink EJ et al (2011) Reduction of *Clostridium sporogenes* spore outgrowth in natural sausage casings using nisin. Food Microbiol 28:974–979

Winkowski K, Crandall AD, Montville TJ (1993) Inhibition of *Listeria monocytogenes* by *Lactobacillus bavaricus* MN in beef systems at refrigeration temperatures. Appl Environ Microbiol 59:2552–2557

Woraprayote W, Kingcha Y, Amonphanpokin P et al (2013) Anti-listeria activity of poly(lactic acid)/sawdust particle biocomposite film impregnated with pediocin PA-1/AcH and its use in raw sliced pork. Int J Food Microbiol 2167:229–235

Ye M, Neetoo H, Chen H (2008) Effectiveness of chitosan-coated plastic films incorporating anti-microbials in inhibition of *Listeria monocytogenes* on cold-smoked salmon. Int J Food Microbiol 127:235–240

Yildirim Z, Yildirim M, Johnson MG (2007) Effects of bifidocin b and actococcin r on the growth of *Listeria monocytogenes* and bacillus cereus on sterile chicken breast. J Food Safety 27:373–385

Zhang J, Liu G, Li P et al (2010) Pentocin 31-1, a novel meat-borne bacteriocin and its application as biopreservative in chill-stored tray-packaged pork meat. Food Control 21:198–202

Zhang S, Mustapha A (1999) Reduction of *Listeria monocytogenes* and *Escherichia coli* O157:H7 numbers on vacuum-packaged fresh beef treated with nisin or nisin combined with EDTA. J Food Prot 62:1123–1127

Zoumpopoulou G, Foligne B, Christodoulou K et al (2008) *Lactobacillus fermentum* ACA-DC 179 displays probiotic potential in vitro and protects against trinitrobenzene sulfonic acid (TNBS)-induced colitis and *Salmonella* infection in murine models. Int J Food Microbiol 121:18–26

Chapter 5
Biopreservation of Milk and Dairy Products

5.1 Application of Bacteriocin Preparations

5.1.1 Raw Milks

Milk may act as vehicle for human pathogenic bacteria (reviewed by Claeys et al. 2013). Pasteurization of milk before human consumption or for the manufacture of dairy products is often required or recommended. Pasteurizarion will decrease the background spoilage microbiota, but it will not yield a sterile product. Some traditional, highly appreciated fermented dairy foods are still made from raw milk, and there is an ongoing debate on the benefits of consuming raw milk versus pasteurized milk (Claeys et al. 2013). According to foodborne disease reports from different industrialized countries, milk and milk products are implicated in 1–5 % of the total bacterial foodborne outbreaks, with 39.1 % attributed to milk, 53.1 % to cheese and 7.8 % to other milk products (De Buyser et al. 2001; Claeys et al. 2013). Bacteriocins seem an attractive approach to improve the safety of milk and dairy products (especially in those made from raw milk), and at the same time may offer some potential technological applications such as in acceleration of cheese ripening (Table 5.1). The antimicrobial effects of bacteriocins and/or their produced strains have been investigated both in raw milks and in several types of dairy products.

Many different bacteriocins preparations have been tested for preservation of milks and dairy products, with the purpose of inactivating foodborne pathogenic or spoilage bacteria. Addition of nisin to raw milks may help solving particular shelf-life problems associated with hot weather temperature and/or long distance transport and inadequate refrigeration systems (Davies and Delves-Broughton 1999; Thomas et al. 2000). The application of nisin in combination with heat treatments decreased the D values of bacteria such as *Bacillus cereus* and *Geobacillus stearothermophilus* and natural microbiota, making it possible to apply milder thermal treatments and at the same time extend the shelf life of milk even under poor refrigeration conditions. Another suggested approach was to use coatings containing

© The Author(s) 2014
A. Gálvez et al., *Food Biopreservation*, SpringerBriefs in Food, Health, and Nutrition, DOI 10.1007/978-1-4939-2029-7_5

Table 5.1 Examples of bacteriocin applications in dairy foods

Bacteriocin treatment	Effect(s)	Reference(s)
Nisin	Prevent proliferation of surviving endospore formers, mainly the gas-producing clostridia and *C. botulinum* in cheeses	Thomas and Delves-Broughton 2001
	Prevent post-process contamination with *L. monocytogenes*	
Nisin and PEF	Increased antimicrobial activity in milks against several bacteria such as *L. monocytogenes, S. aureus, B. cereus* and *E. coli*	Sobrino-López and Martín-Belloso 2008
Nisin and HHP	Increased the inactivation of spoilage bacteria associated with milk	Black et al. 2005
Lacticin 3147	Inactivation of *L. monocytogenes* in reconstituted demineralized whey poder and *S. aureus* in reconstituted skimmed milk	Morgan et al. 2000
Lacticin 3147	Inactivation of *L. monocytogenes* in natural yogurt and in Cottage cheese	Morgan et al. 2001
Enterocin AS-48	Rapid inactivation of *L. monocytogenes* and slower inhibition of *S. aureus* in skim milk	Ananou et al. 2010
Enterocin AS-48 and PEF	Enhanced inactivation of *S. aureus* in skim milk	Sobrino et al. 2009
Pediocin PA-1/AcH	Inhibition of *L. monocytogenes* in several dairy systems (dressed Cottage cheese, half-and-half cream, cheese sauce, and others)	Rodríguez et al. 2002
Lacticin 3147-producing cultures	Inhibition *L. monocytogenes* in Cottage cheese, and on the surface of a mould-ripened cheese and a smear-ripened cheese	O'Sullivan et al. 2006
Enterocin AS-48 producer *E. faecalis* strain	Inhibition of *B. cereus* and *S. aureus* in cheeses and in skim milk	Muñoz et al. 2004, 2007
Lacticin 3147-producing *L. lactis* IFPL 3593	Inhibition of gas formation by *C. tyrobutyricum* and heterofermentative lactobacilli in cheese	Martínez-Cuesta et al. 2010
Bacteriocin producer *L. gasseri* K7	Reduced outgrowth of inoculated *C. tyrobutyricum* and butyric acid formation in the cheeses. Probiotic properties	Bogovič Matijašić et al. 2007
S. macedonicus ACA-DC (producer of macedocin)	Inhibition of gas formation by *C. tyrobutyricum* in cheese	Anastasiou et al. 2009
Lacticin 481-producing culture	Prevention of late-blowing defects due to inoculated *C. beijerinckii*	Garde et al. 2011
Lacticin 3147-producing cultures	Increased generation of 2-methylbutanal with the concomitant enhancement of the cheese aroma	Fernández de Palencia et al. 2004
Lacticin 3147-producing cultures	Inhibition of adventitious non-starter LAB flora during ripening, enhancing the cheese quality	Ryan et al. 2001

immobilized nisin for milk packaging. In experimental trials, a low-density polyethylene film coated with nisin retarded growth of *Micrococcus luteus* as an indicator strain in raw, pasteurized and UHT milk (Mauriello et al. 2005). Similar results were reported with virgin paperboard coated with nisin and/or chitosan in a binder of vinyl acetate–ethylene copolymer during storage of pasteurized milk (Lee et al. 2004), or with cross-linked hydroxypropylmethylcellulose (HPMC) films containing nisin (Sebti et al. 2003).

The antibacterial activity of nisin in milk can be enhanced in combination with other antimicrobials, such as monolaurin, the lactoperoxidase (LPS) system, lysozyme or reuterin (Gálvez et al. 2007), or even by addition of bioactive culture supernatants as reported for *Bacillus licheniformis* ZJU12 (He and Chen 2006). Application of nisin in combination with pulsed-electric fields (PEF) or high hydrostatic pressure (HHP) has been investigated in recent years with the purpose of increasing the efficacy of treatments while having lower impact on the food organoleptic properties and nutritional value (Black et al. 2005; Sobrino-López and Martín-Belloso 2008). The combined application of PEF and nisin was shown to improvethe inactivation of *Listeria monocytogenes, Staphylococcus aureus, B. cereus* and *Escherichia coli* in pumpable substrates such as skim milk, whey, or simulated milk ultrafiltrate media (Calderón-Miranda et al. 1999; Pol et al. 2001; Terebiznik et al. 2002; Sobrino-López et al. 2006; Sobrino-López and Martín-Belloso 2008). Increasing the electric field intensity, the number of pulses and the nisin concentration acted synergistically for inactivation of *Listeria innocua* in skim milk, achieving up to 3.8 log units inactivation after application of a PEF treatment of 32 pulses at 50 kV/cm in combination with 100 IU/ml nisin (Calderón-Miranda et al. 1999). Similar reductions of cell viability were observed on the natural microbiota of raw milk (Smith et al. 2002) and on *S. aureus* inoculated in skim milk (Sobrino-López et al. 2006). Synergy between PEF treatment and nisin may be further enhanced by inclusion of a third hurdle, such as a mild thermal treatment or the addition of other antimicrobials such as carvacrol or lysozyme (Sobrino-López and Martín-Belloso 2008). For example, the combination of PEF treatment (50 pulses at 80 kV/cm and 52 °C) with nisin (38 IU/ml) and lysozyme (1,638 IU/ml) achieved at least 7.0 log reduction of the milk endogenous microbiota (Smith et al. 2002).

Application of combined treatments of nisin and HHP (with or without lysozyme) increased the inactivation of bacteria associated with milk such as *E. coli, Pseudomonas fluorescens, L. innocua,* and *Lactobacillus viridescens* (García-Graells et al. 1999; Black et al. 2005). For example, treatment at 500 MPa for 5 min in combination with nisin (500 IU/ml) completely inactivated *P. fluorescens* and *E. coli* and reduced *L. innocua* by more than 8.3 log cycles, while reductions of cell viability obtained for the single treatments were considerably lower, in the range of 1.5–3.8 log cycles (Black et al. 2005). Milk exerts a protective effect on bacteria against HHP. For that reason, and also to overcome the higher resistance of some bacteria, nisin was combined with other antimicrobials in HHP treatments. For example, the combination of nisin and pediocin PA-1 (500 AU/ml total activity) with HHP treatment (345 MPa, 50 °C, 5 min) reduced viable cell counts of *S. aureus* by more than 8-log cycles, and also avoided growth of possible survivors for at least

30 days of storage (25 °C) of the treated milk samples (Alpas and Bozoglu 2000). In another study, the combination of nisin and lysozyme with HHP improved inactivation of pressure-resistant *E. coli* in skim milk, although the efficacy of the combined treatment decreased as the fat content of milk increased (García-Graells et al. 1999).

Lacticin 3147 is another lactococcal bacteriocin with a high potential for application in the preservation of dairy foods (Ross et al. 1999; O'Sullivan et al. 2002b). The effects of lacticin 3147 preparations on the inactivation of *L. monocytogenes* in reconstituted demineralized whey poder and *S. aureus* in reconstituted skimmed milk. Lacticin addition at 10,000 AU ml achieved a 0.7 log reduction in viability for both bacteria after 30 min incubation, while 20,000 AU achieved a 2.1 log reduction of the listeria population (Morgan et al. 2000). Microbial inactivation increased greatly when lacticin preparations were tested in combination with HHP treatments. The combination of 250 MPa and lacticin 3147 resulted in more than 6 logs of kill for both bacteria. It was also shown that the antibacterial activity in concentrated preparations of this lacticin against *L. monocytogenes* and *S. aureus* could be enhanced in combination with HHP treatment (150–275 MPa), thereby reducing the amounts of bacteriocin required (Morgan et al. 2000).

Only a few enterocins have been investigated in milks. For example, enterocin CCM 4231 was able to inhibit *L. monocytogenes* and *S. aureus* in milk (Lauková et al. 1999). Enterocin AS-48 (50 µg/ml) added to skim cow milk partially reduced the population of inoculated *S. aureus* during the first 10 h of incubation at 28 °C, but did not avoid overgrowth at 24 h (Muñoz et al. 2007). However, a lower bacteriocin dose (20 µg/ml) in combination with a mild heat treatment (65 °C, 5 min) was highly effective in reducing the population of staphylococci in the milk below detectable levels within the first 8 h of incubation and also to avoid overgrowth of staphylococci at 24 h. *Enterococcus faecalis* INIA 4 produces enterocin 4, which is identical to enterocin AS-48. In bacteriocin filtrates prepared by cultivation of this strain on skimmed ewe's milk, the levels of inoculated *L. monocytogenes* Ohio and Scott A strains were reduced by 3.23 and 2.13 log CFU/ml respectively, after 24 h incubation at 30 °C. Reductions obtained under similar conditions for a collection of *L. monocytogenes* strains isolated from dairy environments were in the range of 0.52–3.48 log cycles (Rodríguez et al. 1997). Enterocin AS-48 can now be produced as a dry powder based on a whey permeate that is essentially free of dairy allergens. Addition of this bacteriocin preparation to skim milk, rapidly inactivated the inoculated *L. monocytogenes* cells and progressively reduced the counts of *S. aureus* (Ananou et al. 2010).

Microbial inactivation in skim milk improved when bacteriocins were tested in combination with PEF. The combinations of enterocin AS-48 with or without nisin and PEF treatment increased the inactivation of *S. aureus* (Sobrino et al. 2009). A maximum of 4.5 logs reduction was achieved for AS-48 (28 AU/ml) and PEF (35 kV/cm). This reduction factor increased up to 6 logs when 28 AU/ml of AS-48 and 20 IU/ml of nisin were added to the milk before PEF treatment. The combined treatment extended the shelf life of milk by at least 1 week compared to a conventional pasteurisation treatment.

5.1.2 Processed Milk Products

Nisin is widely used in the dairy industry for inhibition of gas blowing defect in cheeses caused by *Clostridium tyrobutyricum*, but also in processed cheeses and cheese products to inhibit *Clostridium botulinum*, and to prevent growth of post-process contaminating bacteria such as *L. monocytogenes* (Davies and Delves-Broughton 1999; Thomas et al. 2000; Thomas and Delves-Broughton 2001; Deegan et al. 2006; Sobrino-López and Martín-Belloso 2008) (Table 5.1). It is also used in many other pasteurised dairy products, such as chilled desserts, flavoured milk, clotted cream, or canned evaporated milks (Thomas et al. 2000). For example, the addition of nisin powder to milk in the production of cheese made without a starter culture can control microbial contamination (Sobrino-López and Martín-Belloso 2008). Addition of nisin at 100 or 500 mg/kg suppressed total plate and anaerobic spore counts in processed cheese during 3 months of storage at 5 or 21 °C, and even the growth of *G. stearothermophilus*, *B. cereus* and *Bacillus subtilis* were inhibited by 5 mg/kg nisin (Plockova et al. 1996). In Ricotta-type cheese, addition of nisin (2.5 mg/l) inhibited the growth of *L. monocytogenes* for more than 8 weeks, while cheese made without nisin contained unsafe levels of the bacteria within 1–2 weeks. In addition, there was a high level of retention of nisin activity in the cheese after 10 weeks of incubation at 6–8 °C, with only 10–32 % loss of antibacterial activity (Davies et al. 1997).

Nisin addition in combination with HHP could be useful for inactivation of endospores and mesophilic bacteria in cheese (Capellas et al. 2000; López-Pedemonte et al. 2003; Arqués et al. 2005a). For example, the combination of nisin with HHP strongly reduced the counts of *B. cereus* spores in a traditional cheese curd. Since bacterial endospores are pressure-resistant, two HHP cycles were applied, the first one to induce endospore germination and the second one to destroy the vegetative cells (López-Pedemonte et al. 2003). The authors of this study concluded that the combined treatment could improve the microbial stability and safety of cheeses, especially those made from unpasteurised milk, such as many traditional cheeses, decreasing the restrictions that are currently imposed on the commercialisation of such cheeses.

Due to its non-specific inhibitory activity against Gram-positive bacteria, nisin may interfere with growth of starter cultures cheese fermentation and have detrimental effects on acidification and/or aroma formation. In order to solve this limitation and also to enhance nisin stability, nisin Z was encapsulated in liposomes (Benech et al. 2002a, b). Addition of liposome preparations was shown to inhibit listeria in Cheddar cheese (Benech et al. 2002a, b). In another study, nisin encapsulated in soybean phosphatidylcholine nanovesicles provided best results compared to added free nisin in the control of *L. monocytogenes* in Minas frescal cheese stored at 7 °C, keeping the concentrations of listeriae lower than the untreated controls for at least 21 days (Malheiros et al. 2012). Some growth inhibition was also obtained with an encapsulated BLIS P34 (derived from *Bacillus* sp. P34) in parallel experiments.

The authors of this study concluded that encapsulation of bacteriocins in liposomes of partially purified soybean phosphatidylcholine may be a promising technology for the control of foodborne pathogens in cheeses.

Processing of dairy foods, such as cheese slicing, can be a critical point for bacterial contamination. For application on sliced cheeses, nisin immobilized in polyethylene/polyamide packagings was shown to reduce the population of LAB, *L. innocua* and *S. aureus* on the cheese slices packaged with the activated interleaves (Scannell et al. 2000). Also, when nisin immobilized in sodium caseinate films was tested against *Listeria* inoculated on the surface and in depth on mini red Babybel soft cheese, the presence of the active film resulted in a 1.1 log CFU/g reduction in *L. innocua* counts on the cheese surface after 1 week of storage at 4 °C as compared to control samples (Cao-Hoang et al. 2010). Inactivation rates decreased as depth of inoculation increased, e.g. 1.1, 0.9 and 0.25 log CFU/g for distances from the contact surface of 1, 2, and 3 mm, respectively, reflecting the nisin diffusion gradient. The study concluded that nisin immobilized in sodium caseinate films a promising method to overcome problems associated with post-process contamination, thereby extending the shelf life and possibly enhancing the microbial safety of cheeses.

Added lacticin 3147 powder rapidly inactivated *L. monocytogenes* and reduced *S. aureus* viable cell counts in an infant milk formulation, and was highly effective against *L. monocytogenes* in natural yogurt and in Cottage cheese (Morgan et al. 2001). Addition of 10 % lacticin 3147 powder reduced the concentration of viable *Listeria* below detectable levels in yogurt within 60 min or killed 85 % of cells in cottage cheese within 120 min (Morgan et al. 2001). However, optimisation of lacticin 3147 powder to increase specific activity may be necessary in order to decrease the amount of added powder required for an effective microbial inhibition.

Several enterococal bacteriocins have been tested for preservation of dairy foods (Giraffa 1995; Foulquié Moreno et al. 2006; Gálvez et al. 2008). Enterocins CCM 4231, CRL35, or AS-48 can reduce the levels of *L. monocytogenes, S. aureus* or *B. cereus* in dairy products. A concentrated enterocin CRL35 preparation added to goat cheese (10,400 AU/ml) reduced the population of *L. monocytogenes* by 9 log units by the end of ripening period without affecting the cheese quality (Farías et al. 1999). Lauková et al. (1999) reported that addition of enterocin CCM 4231 (3,200 AU/ml) in yogurt milk inoculated with *L. monocytogenes* reduced the levels of listeria after 24 h incubation at 30 °C by about 2 log CFU/ml. Similarly, enterocin addition (3,200 AU/ml) to skim milk decreased the viable counts of *S. aureus* from 10 log CFU/ml to 2 log CFU/ml after 24 h incubation of milk at 27 °C. This bacteriocin also reduced the levels of *Listeria* in "bryndza" (a traditional Slovak soft cheese from sheep milk) and Saint-Paulin cheese (Lauková et al. 2001; Lauková and Czikková 2001) but it did not achieve complete elimination of the bacteria. For example, addition of enterocin CCM 4231 during Saint-Paulin cheese preparation (3,200 AU/ml) reduced the population of inoculated *L. monocytogenes* by almost 5 log cycles for up to 1 week, followed by proliferation of the listeria afterwards(Lauková et al. 2001).

Bcteriocins from enterococci have been used to prepare activated films or coatings for use in the cheese industry. In one study, Iseppi et al. (2008) tested the antilisterial effects of film packagings activated with enterocin 416K1 on fresh cheese surfaces. In the fresh soft cheese samples packed in enterocin-activated film, the listeria counts were lower than controls by about 1 log unit up to 28 days for samples stored at 4 °C and for up to 7 days for samples stored at 22 °C. This approach for application of bacteriocins not only reduces the levels of listeria on cheese surfaces but is also an effective barrier against cross-contamination of the cheeses.

Enterocin AS-48 has been suggested for biopreservation of prepared dishes and desserts containing milk. In boiled rice and in a commercial rice-based infant formula dissolved in whole milk inoculated with vegetative cells or with endospores of *B. cereus,* enterocin AS-48 (20–35 µg/g) reduced viable cell counts below detectable levels during storage for up to 15 days in a temperature range of 6–37 °C and prevented enterotoxin production (Grande et al. 2006a). Bacteriocin activity was improved by adding sodium lactate, decreasing the effective bacteriocin concentration to 8–16 µg/g. Although the bacterial endospores were resistant to this bacteriocin, application of AS-48 in combination with heat treatments decreased the thermal death *D* values for endospores (Grande et al. 2006b). In desserts and bakery ingredients, the bactericidal effect of AS-48 on *S. aureus, B. cereus* and *L. monocytogenes* depended on the food substrate and the target bacteria (Martínez Viedma et al. 2009a, b). The lowest and highest efficacies were always detected in soy-based desserts and in gelatin pudding, respectively. *L. monocytogenes* was completely inactivated by bacteriocin concentrations in the range of 5–25 µg/g, depending on the substrate, and *B. cereus* was inactivated in a range of 15–50 µg/g. Bacteriocin addition to gelatin pudding prevented the production of proteases by *B. cereus* and the consequent gelatin liquefaction. Inactivation of *S. aureus* required a higher bacteriocin concentration (50 µg/g) and also a lower population density of staphylococci, not higher than 5 log CFU/g. The bacteriocin also showed a variable degree of activity against *S. aureus* in substrates like pumpkin confiture, diluted almond cream or liquid caramel, but was ineffective in vanilla or chocolate creams (Martínez Viedma et al. 2009). In chocolate cream, where higher bacteriocin concentrations were required because of interaction of the bacteriocin with the food substrate, antimicrobial activity increased markedly when AS-48 was tested in combination with eugenol, 2-nitropropanol or Nisaplin.

Pediocin PA-1/AcH preparations are interesting for application in dairy products due to the bacteriocin antilisterial activity, stability in aqueous solutions at ambient temperature and also during freezing and heating, and wide pH range for activity (Nes et al. 1996; Rodríguez et al. 2002). The commercial preparations containing pediocin in the form of Alta™ products can be used as ingredients in dairy foods. Several studies have shown that added pediocin PA-1/AcH is effective in reducing the levels of *L. monocytogenes* in several types of dairy products such as dressed Cottage cheese, half-and-half cream, and cheese sauce (reviewed by Rodríguez et al. 2002).

Other bacteriocins of interest in preservation of dairy foods are the propionicins. Propionibacteria are used in some dairy fermentations and may produce bacteriocins with broad inhibitory spectra (Holo et al. 2002). Microgard™ is a commercial preparation containing an antimicrobial peptide produced by *Propionibacterium freudenreichii* ssp. *shermanii* (Weber and Broich 1986), which is approved in certain countries for commercial use as an ingredient mainly in dairy products such as Cottage cheese and yogurt. Bacteriocins produced by *Propionibacterium jensenii* P126 and P1264 strains have been patented as anti-bacterial agents for controlling the growth of certain lactic acid bacteria. These bacteriocins could be particularly useful in controlling the over-acidification of yogurt to decrease the sour taste often found in this product.

Some bacteriocins from bacteria not associated with milk fermentations been investigated for application in dairy foods. Variacin in the form of a dry milk-based ingredient inhibited the proliferation of *B. cereus* in chilled dairy products, vanilla and chocolate desserts (Mollet et al. 2004). There is also a growing interest in exploitation of bacteriocins from bacilli. Cerein 8A is an antimicrobial peptide produced by the soil isolate *B. cereus* 8A, with bactericidal activity towards *L. monocytogenes* and *B. cereus* (Bizani et al. 2005). Cerein addition inhibited growth of *L. monoytogenes* in milk and on the surface of Minas-type cheese during refrigeration storage, suggesting its potential use as biopreservative in dairy products.

5.2   Application of Bacteriocin-Producing Strains

5.2.1   *Inhibition of Foodborne Pathogens*

L. monocytogenes is considered the main foodborne pathogen of concern in cheese and dairy products. Therefore, many different studies have focused on the application of antilisterial starter or adjunct cultures for inhibition of this bacterium (Table 5.1). Nisin-producing lactococcal strains inhibit *L. monocytogenes* in several types of cheeses such as Cottage, Camembert or Manchego cheese made from raw milk. They can also reduce *S. aureus* viable counts (Deegan et al. 2006; Gálvez et al. 2008), but often lack the technological properties required for cheese making such as fast acidification capacity and proteolytic activity. For this reason, they should be recommended as adjunct cultures in combination with suitable nisin-resistant strains as the primary starters. Nevertheless, a comparative study on performance of bacteriocin-producing *Lactococcus lactis* strains selected on technological criteria for Cottage cheese fermentation (one nisin Z producer, one nisin A producer and two lacticin 481 producers) established that the nisin A producing *L. lactis* 40FEL3, and to a lesser extent the lacticin 481 producers 32FL1 and 32FL3, successfully controlled the growth of the pathogen during the manufacture and storage of Cottage cheese (Dal Bello et al. 2011).

The lactococci can also produce other bacteriocins (such as lacticin 3147 or lacticin 481) in fermented dairy products (Guinane et al. 2005). Lacticin 3147-producing starter cultures have been tested to control the non-starter lactic acid bacteria population in Cheddar cheese (Ryan et al. 1996). In order to develop suitable starters, the plasmid coding for lacticin 3147 production was transferred to suitable recipient lactococci. Food-grade lactococcal starters which produce the lantibiotics lacticin 3147 and lacticin 481 have also been reported (O'Sullivan et al. 2003). Lacticin production by lacticin 3147 modified starters successfully inhibited *L. monocytogenes* in Cottage cheese, in semi-hard raw-milk cheeses and on the surface of a mould-ripened cheese and smear-ripened cheese (O'Sullivan et al. 2006). In Cottage cheese inoculated with a *L.lactis* transconjugant strain, the bacteriocin concentration in the curd reached 2,560 AU/ml, and bacteriocin activity could be detected throughout the 1 week storage period. In cottage cheese samples held at 4 °C, there was at least a 99.9 % reduction in the numbers of *L. monocytogenes* Scott A in the bacteriocin-containing cheese within 5 days, whereas in the control cheeses, numbers remained essentially unchanged. At higher storage temperatures, the kill rate was more rapid (McAuliffe et al. 1999). In a smear-ripened cheese, applications of a live lacticin 3147-producing culture on a cheese surface containing *Listeria,* the viable cell concentrations of the pathogen were found to be up to 100-fold lower than in the cheese treated with a bac- *L. lactis* strain as control. The lactococci have also been tested for heterologous production of other bacteriocins such as enterocin A. The resulting starter derivative successfully controlled the levels of *L. monocytogenes* during Cottage cheese fermentation (Liu et al. 2008).

Enterococci are well adapted to grow in milk, and can produce sufficient bacteriocin amounts in milk substrates as to inhibit pathogenic bacteria such as *L.monocytogenes* and others. For example, *E. faecalis* EJ97 produced enterocin EJ97 during cocultivation in half-skimmed milk, although its capacity to control *L. monocytogenes* was limited to listerial populations of low densities ($\leq 10^3$ CFU/ ml; García et al. 2004). *Enterococcus faecium* strain F58, isolated from Moroccan jben goat's cheese, was also able to produce bacteriocin in milk and to achieve a partial inhibition of *L. monocytogenes* during cocultivation in goat milk (Achemchem et al. 2006). In addition, pre-cultivation of strain F58 in milk for 12 h before inoculation of the listeria (at 3 log CFU/ml) produced enough bacteriocin to completely inactivated the inoculated listeria during further incubation. The enterocin AS-48 producer strain *E. faecalis* A-48-32 was able to produce enough bacteriocin in nonfat cow milk to reduce the population of co-inoculated *B. cereus* below detectable levels after 72 h of cocultivation at 30 °C (Muñoz et al. 2004). In cocultures done with skim milk, this strain was also able to control *S. aureus* (Muñoz et al. 2007).

Enterococci are very often found as part of the adventitious microbiota in fermented foods, including many traditional cheeses, and exhibit many biochemical properties of technological interest in dairy fermentations such as production of organic acids and acidification, proteolytic and peptidolytic activities, lipolytic and esterase activities, and citrate and pyruvate metabolism, together with their capacity to produce bacteriocins (Giraffa 2003). Bacteriocin-producing enterococci have been investigated as adjunct cultures for cheese making because of their robustness, natural presence in cheeses, and production of several bacteriocins with strong anti-listerial activity (Giraffa 1995; Foulquié Moreno et al. 2003; Franz et al. 2007; Gálvez et al. 2008). Inoculation of Jben goats' milk cheese with bacteriocinogenic strain *E. faecium* F58 as an adjunct culture, caused a sharp decrease in the concentration of viable *L. monocytogenes*, which were undetectable after 1 week of cheese storage at 22 °C (Achemchem et al. 2006). The bacteriocinogenic strain of *E. faecium* 7C5 (which produces enterocin AS-48), added as an adjunct with a thermophilic culture in soft cheese, led to complete death of *L. monocytogenes* and *L. innocua* without altering the acidifying activity of the starter culture (Giraffa et al. 1995b). When tested in cheeses, other strains producing enterocin AS-48 showed strong inhibition of *L. monocytogenes*, as well as *B. cereus* and *S. aureus* (Núñez et al. 1997; Muñoz et al. 2004, 2007). During the manufacture of Manchego cheese made from raw ewe's milk, *E. faecalis* INIA 4 was able to compete with cheese microbiota and produce enterocin (Núñez et al. 1997). In cheeses inoculated with INIA 4 strain, *L. monocytogenes* Ohio counts were reduced below 1 log CFU/g in the ripened cheese for up to 60 days. However, *L. monocytogenes* Scott A did not seem to be affected by the inoculated enterocin-producer in the cheese. In a separate study, when the enterocin AS-48 producer strain *E. faecalis* A-48-32 was co-inoculated with *B. cereus* during cheese manufacture, viable cell counts of the bacilli were 5.6 log units lower than controls after 30 days of ripening (Muñoz et al. 2004). The efficacy of strain A-48-32 against *S. aureus* was lower compared to *B. cereus*, but staphylococci counts in treated cheeses remained at least 1 log CFU/g below controls throughout at least 1 month storage (Muñoz et al. 2007).

Enterococcus faecium RZS C5 (a natural cheese isolate carrying the structural genes for enterocins A, B and P) was reported to be effective as an anti-listeria bacteriocin-producing co-culture in Cheddar cheese manufacture. The strains *Enterococcus mundtii* CRL35 and *E. faecium* ST88Ch isolated from cheeses were tested for their capability to control growth of *L. monocytogenes* 426 in experimentally contaminated fresh Minas cheese during refrigerated storage (Vera Pingitore et al. 2012). Growth of *L. monocytogenes* 426 was inhibited in cheeses containing *E. mundtii* CRL35 up to 12 days at 8 °C, stressing the potential of this strain for application in Minas cheese. However, *E. faecium* ST88Ch was less effective in the control of listeriae.

Strains of enterococci and lactococci producing bacteriocins (such as enterocins I, TAB 7, TAB 57, AS-48, nisin A, nisin Z and lacticin 481) have been tested in combination with HHP treatments with the aim to improve the safety of cheeses made from raw milk. Inoculation of milk with bacteriocinogenic strains before cheese making followed by application of HHP treatment to the cheeses was reported to increase the bactericidal activity against *L. monocytogenes*, *S. aureus* and *E. coli* O157:H7. For example, for 300 MPa treatment (10 min), counts of *L. monocytogenes* were always lower in the cheeses inoculated with the bacteriocin-producing enterococci both on day one after treatment and also after 60 days of ripening (at 12 °C under vacuum). For *S. aureus*, inoculation of cheeses with bacteriocin-producing enterococci also improved the lethal effects of HHP treatments (Arqués et al. 2005b). When similar treatments (300 MPa, 10 min; 500 MPa, 5 min) were applied to cheeses challenged with *E. coli* O157:H7, the reductions obtained for the treatment at 300 MPa were in the range of 0.7–2 log CFU/g on day one after treatment and 0.2–1.2 log CFU/g after 60 days as compared to the controls not inoculated with bacteriocin producers (Rodríguez et al. 2005). For 500 MPa, greatest differences were observed at day one after treatment, with in which no viable cells were detected in any of the cheese samples inoculated with bacteriocin producers. The authors concluded that the application of reduced pressures combined with bacteriocin-producing enterococci can improve cheese safety while decreasing the deleterious effects on cheese quality caused by HHP at higher pressures.

During the manufacture of smear-ripened cheeses, smear operations increase the risk for surface contamination and cross-contamination with *L. monocytogenes*. In the process of Taleggio smear cheese making, inoculated *E. faecium* 7C5 was reported to produce bacteriocin (Giraffa et al. 1995a) and to inhibit the growth of *L. monocytogenes* Ohio on the cheese surface (Giraffa and Carminati 1997). When *E. faecium* WHE 81, a multi-bacteriocin producer isolated from Munster cheese, was inoculated on the cheese surface during smearing operation, further inoculation of the cheeses with a low level inoculum of *L. monocytogenes* (50 CFU/g) resulted in suppression of the listeria or its complete growth inhibition compared with the controls inoculated with a bac⁻ strain (Izquierdo et al. 2009). The inoculated enterococci had no detrimental effects on the on pH, fungal flora or pigmented bacteria in the cheese rind during ripening.

Thermophilic streptococci are important as dairy starters used in large scale in the production of yogurt and certain cheese varieties. The bacteriocin-producer strain *Streptococcus salivarius* subsp. *thermophilus* B was tested as a thermophilic starter in yogurt to control *L. monocytogenes* and *S. aureus*. The counts of *Listeria* were reduced below detectable levels, but the staphylococci survived in the produced yogurt. Use of the bac + starter was reported to extend the product shelf-life by 5 days (Benkerroum et al. 2002).

Pediococci are not well adapted to dairy substrates, due to their lack or very slow lactose fermentation activity (Papagianni and Anastasiadou 2009). However, some strains such as *Pediococcus acidilactici* NRRL-B-18925 are particularly effective in producing bacteriocin in milk based media. Since pediocin PA-1/AcH is not inhibitory to bacterial species employed as yogurt starters (Gonzalez and Kunka 1987) there is a great interest for application of producer strains in developing more naturally preserved yogurts and to avoid proliferation of cross-contaminating pathogens during yogurt processing. Vedamuthu Ebenezer (1995) patented a method for producing a yogurt product which contains bacteriocin active against undesirable microbiota. The yogurt product can be dried for use in various foods. Pediocin production in milk has been reported in coculture with yogurt starter cultures, at the expense of the excess sugar released from lactose hydrolysis by the starters (Somkuti and Steinberg 2010). While no bacteriocin production was detected when *P. acidilactici* was inoculated into milk as a monoculture, when grown in combination with the yogurt starter cultures *Streptococcus thermophilus* and *Lactobacillus delbrueckii* ssp. *bulgaricus*, pediocin concentration reached 3,200–6,400 units/ml after 8 h of incubation.

The development of pediocin-producing genetically-engineered *L. lactis* strains can be another approach to solve the problems associated with the use of pediococci in dairy substrates. Pediocin-producing lactococci have shown significant potential for inhibition of foodborne pathogens in cheeses. Examples are recombinant *L. lactis* MM217 as a pediocin-producing starter culture in Cheddar cheese (Buyong et al.

1998) or strains of *L. lactis* ESI 153 and *L. lactis* ESI 515, isolated from hand-made raw milk cheese and transformed into pediocin producers (Reviriego et al. 2005). The pediocin-producing transformants reduced viable counts of *L. monocytogenes* in cheese below 50 or 25 CFU/g at the end of the ripening period.

Bacteriocin-producing lactobacilli have also been suggested for preservation of dairy foods. *Lactobacillus plantarum* WHE 92 is a spontaneous pediocin producer that grows well and produces satisfactory pediocin concentrations in Munster cheese (Ennahar et al. 1996). *L. plantarum* LMG P-26358 isolated from a soft French artisanal cheese produces plantaricin 423, which has strong anti-*Listeria* activity. This relatively narrow spectrum bacteriocin also exhibited antimicrobial activity against species of enterococci, but did not inhibit dairy starters including lactococci and lactobacilli. A strong listericidal effect was detected in cheeses made with *L. plantarum* LMG P-26358 as an adjunct culture in combination with a nisin producer, and the single inoculation with LMG P-26358 performed even better than the single nisin producer. Combination of strain LMG P-26358 as adjunct culture with nisin-producing cultures may be an effective strategy to improve the safety and quality of dairy products (Mills et al. 2011). Another interesting example is the human isolate probiotic strain *Lactobacillus gasseri* K7, producer of bacteriocins with wide range of inhibition (Čanžek Majhenič et al. 2003). Application of bacteriocin-producing probiotic strains could be exploited with the purpose of improving food safety and quality and at the same time providing health benefits.

5.2.2 Inhibition of Bacteria Producing Gas-Blowing Defects in Cheeses

Gas production is an undesirable defect in most cheeses, and may be caused by outgrowth of *C. tyrobutyricum* (and other clostridia such as *Clostridium sporogenes, Clostridium beijerinckii* or *Clostridium butyricum*) spores surviving heat treatments applied to milk before processing (Cocolin et al. 2004; Le Bourhis et al. 2007), and also by some heterofermentative LAB. Species of *Clostridium* can ferment lactic acid with production of butyric acid, acetic acid, carbon dioxide and hydrogen. Butyric acid fermentation (also known as late blowing defect) is one of the major causes of spoilage in semi-hard and hard ripened cheeses. This fermentation originates texture and flavor defects in the cheeses, causing important economic losses in the cheese industry (McSweeney and Fox 2004). Bactofugation, milk ultrafiltration, and addition of nitrite or lysozyme are often applied to prevent butyric acid fermentation in cheeses, but also the inoculation of milk with bacteriocinogenic lactic acid bacteria (LAB) in cheese manufacture can provide satisfactory results (Table 5.1). Application of nisin-producing starter cultures to prevent gas-blowing defects in cheese was proposed as early as 1951. Nisin-producing strains have been used for developing a starter culture system for manufacture of Cheddar cheese (Roberts et al. 1992) and Gouda cheese (Bouksaim et al. 2000), among others. Strains producing the natural variant nisin Z have also shown to reduce the levels of *C. tyrobutyricum* in cheese. For example, *L. lactis* ssp. *lactis* IPLA 729, a nisin Z producer isolated from raw milk cheese, was able to produce nisin Z in semi-hard Vidiago cheese. Nisin Z activity reached a concentration of 1,600 AU/ml in 1-day cheeses and this level was maintained until 15 days of ripening. The produced nisin reduced the levels of a *C. tyrobutyricum* spoilage strain inoculated in the cheeses by ca. 3 log cycles, while considerable bacterial growth was observed in control cheeses inoculated with a commercial starter culture and supplemented with nitrate (Rilla et al. 2003). In another study, when Manchego cheese artificially contaminated with endospores of *C. beijerinckii* was inoculated with a *L. lactis* strain producing nisin and lacticin 481 as a starter, no late-blowing defects were observed after 120 days of ripening, and the concentrations of lactic acid and volatile compounds were similar to control cheese (Garde et al. 2011).

Lacticin 3147-producing lactococci have been shown to inhibit *C. tyrobutyricum* spores and prevent late blowing in semi-hard cheeses, and also demonstrated a considerable inhibition of heterofermentative lactobacilli and their associated blowing defects in cheese (Martínez-Cuesta et al. 2010). The authors suggested that application of lacticin producers in cheese manufacture is a promising alternative to the addition of lysozyme, given the increasing concerns about the potential allergenicity of this additive in egg allergic consumers.

Thermophilic streptococci also have a potential for inhibition of *C. tyrobutyricum* in some cheeses. The bacteriocin thermophilin from *S. thermophilus* ST580 is active against *C. tyrobutyricum*, but not against the thermophilic lactobacilli used as starters. Curds made with strain ST580 and inoculated with *C. tyrobutyricum* endospores showed no gas production for up to 20 days (Mathot et al. 2003). This strain could be included in thermophilic starters for hard cheese making. *Streptococcus macedonicus* ACA-DC 198, which produces the food-grade lantibiotic macedocin, could also be employed to inhibit gas formation in cheese (De Vuyst and Tsakalidou 2008). When tested as an adjunct culture in Kasseri cheese production, this strain produced macedocin in the cheese and inhibited outgrowth of *C. tyrobutyricum* spores during the cheese production and ripening (Anastasiou et al. 2009).

Among lactobacilli, the bacteriocin producer strain *L. gasseri* K7 was able to survive during semi-hard-type cheese manufacturing, and reduced outgrowth of *C. tyrobutyricum* and butyric acid formation in the cheeses (Bogovič Matijašić et al. 2007).

5.2.3 Improving Cheese Quality

Most bacteriocins act on the bacterial cytoplasmic membrane, modifying cell permeability. A secondary effect of bacteriocins is the induction of cell lysis, as a result of deregulation of cell wall autolysins (Gálvez et al. 1990). This effect was further observed on dairy starter cultures, and led to more detailed studies on the potential applications for the release of bacterial intracellular enzymes (such as lipases, proteases, peptidases, and amino acid-converting enzymes) of technological relevance in cheese ripening (Lortal and Chapot-Chartier 2005; Peláez and Requena 2005; Deegan et al. 2006).

In one study, *L. lactis* DPC3286 (producing lactococcins A, B, and M) was tested as an adjunct starter in cheddar cheese manufacture to induce lysis of a sensitive acidifying starter strain (Morgan et al. 1997) or in combination with a bacteriocin-resistant starter. In experimental cheeses with the bacteriocin-resistant starter, the levels of free amino acids increased together with a greater release of the intracellular enzyme lactate dehydrogenase (LDH), and the cheeses showed lower bitterness compared to controls (Morgan et al. 2002). Because lactococcins A, B, and M have a narrow spectrum of activity limited to lactococci, their producer strains could be applied specifically for acceleration of cheese ripening (Ross et al. 1999). Another study reported that addition of the nisin Z-producing strain *L. lactis* ssp. *lactis* biovar *diacetylactis* UL719 during cheddar cheese making increased lipolysis and proteolysis, as well as the formation of hydrophilic and hydrophobic peptides, and enhanced the sensory characteristics of cheese (Benech et al. 2003). Strain UL719 was also tested for acceleration of autolysis of an adjunct *L. delbrueckii* subsp. *bulgaricus* strain, with the result of increasing cheese proteolysis and improving the cheese texture (Sallami et al. 2004). An interesting observation was that bacteriocin production by strain UL719 in cocultures with nisin-sensitive starters *Lactobacillus rhamnosus* RW-9595M and *L. lactis* subsp. *cremoris* was stimulated by 3.1- to 4.6-fold. This stimulation was attributed to the high proteolytic activity of *L. cremoris* and to the release of intracellular nutrients due to autolysis and nisin Z-induced lysis (Grattepanche et al. 2007).

Production of lacticin 481 by *L. lactis* subsp. *lactis* strain DPC5552 induced release of the intracellular enzymes LDH and postproline dipeptidyl aminopeptidase by starter strain *L. lactis* HP without completely inhibiting its growth (O'Sullivan et al. 2002a). Bacteriocin production also induced the release of elevated levels of LDH from the starter without severely compromising its acid-producing capabilities in a cheddar cheese–making trial. Further studies indicated that lacticin 481 was also able to accelerate starter cell lysis (O'Sullivan et al. 2003), and that lacticin 481-producing cultures promoted early lysis of *Lactobacillus helveticus* cells in Hispánico cheese and increased the proteolytic activity (Garde et al. 2006). During the Hispánico cheese–making process, inoculation of milk with strain *L. lactis* subsp. *lactis* INIA 415 (harboring the structural genes of lacticin 481 and nisin Z production) promoted early lysis of mesophilic and thermophilic starter bacteria and increased extracellular aminopeptidase activity (Garde et al. 2002). It also lowered the ratio of hydrophobic-to-hydrophilic peptides, increased the free amino acid content (Avila et al. 2006), and enhanced the formation of several volatile compounds of relevance to the odor and aroma of the cheese, such as hexanal, 2-methyl-1-propanol, 3-methyl-1-butanol, acetone, 2-pentanone, 2-hexanone, and 2-heptanone, but decreased the formation of acetaldehyde, ethanol, 3-methyl-3-buten-1-ol, 3-methyl-2-buten-1-ol, ethyl acetate, ethyl butanoate, ethyl hexanoate, 2-butanone, 2,3-butanedione, 2,3-pentanedione, and 3-hydroxy-2-butanone (Garde et al. 2005).

Lacticin 3147 production in cheese also induced bacterial lysis of cheese starters, increased cheese proteolysis and facilitated the access of bacterial amino-acid converting enzymes to amino acids (Deegan et al. 2006). The accelerated starter cell lysis enhanced reactions such as isoleucine transamination, increasing the formation of alpha-keto-beta-methyl-n-valeric acid and 2-hydroxy-3-methyl-valeric acid and cheese aroma intensity due to the higher 2-methylbutanal formation. Optimisation of aroma production could be achieved by selective combination of starters, such as a lacticin 3147-producing *L. lactis* transformant in combination with adjunct cultures producing aminotransferase and a-keto acid decarboxylase activities (Fernández de Palencia et al. 2004).

The application of bacteriocin-producing adjunct cultures to accelerate cheese ripening can be a cheaper approach compared to the addition of exogenous lytic enzymes (Papagianni and Anastasiadou 2009). Application of bacteriocins and/or their producer strains in the development of stabilised cheese flavouring systems has been patented (Dias et al. 2009). Using microbial cells with high aminopeptidase activity in combination with an antimicrobial that can permeabilize the cells can decrease the levels of aminopeptidases that need to be added to the matrix, thereby increasing efficiency. Cheeses produced at the expense of enzymes released form the bacterial cells have a more rounded flavour (Dias et al. 2009). An adjunct *P. acidilactici* culture that accelerates and enhances flavour formation in Cheddar and semi-hard cheeses due to the production of bacteriocins has been marketed by Danisco (CHOOZIT™ Lyo. Flav 43).

Another suggested application of bacteriocin-producing cultures is the inactivation of adventitious non-starter LAB (NSLAB) microbiota during cheese ripening. NSLAB can proliferate during ripening and often tend to become the dominant microbiota in the cheese. The precise role of NSLAB strains in flavor development remains unclear, but they certainly contribute in the quality of many cheeses (Fox et al. 1998). Control of NSLAB is still a pending issue in dairy industries. Growth of NSLAB may induce batch to batch variations in the sensory quality of cheese and cause defects such as the formation of calcium lactate crystals (due to racemation of L-lactate to D-lactate), slit formation and off-flavour development, but they may also exert beneficial effects on the cheeses. Application of lacticin 3147-producing starters has been proposed as a way to enhance cheese quality through inhibition of adventitious NSLAB microbiota during ripening (Ryan et al. 2001; Deegan et al. 2006). Co-inoculation of a lacticin-3147 adapted *Lactobacillus paracasei* subsp. *paracasei* adventitious strain (isolated from a well-flavored, commercial Cheddar cheese) with a lacticin-3147 producing starter culture allowed a better control of NSLAB microbiota during ripening. By using randomly amplified polymorphic DNA-PCR, it was demonstrated that the resistant adjunct strain comprised the dominant microflora in the test cheeses during ripening (Ryan et al. 2001). During Cheddar cheese manufacture, ripening can be accelerated by increasing the temperature from 7 to 12 °C, but this also results in a higher risk of spoilage due to a more rapid proliferation of NSLAB. Inoculation with a lacticin 3147-producing strain allowed a better control of NSLAB during cheese ripening at elevated temperature. Lacticin 481 production has also shown to reduce the concentrations of

NSLAB in cheese by 4 to 2 orders of magnitude during ripening. Bacteriocin production in the cheese resulted in selection of NSLAB that were much more resistant to the bacteriocin than isolates from control cheeses. Therefore, it would be possible to select bacteriocin-resistant strains that do not have negative effects on cheese ripening as the predominant NSLAB.

References

Achemchem F, Abrini J, Martinez-Bueno M et al (2006) Control of *Listeria monocytogenes* in goat's milk and goat's jben by the bacteriocinogenic *Enterococcus faecium* F58 strain. J Food Protect 69:2370–2376

Alpas H, Bozoglu F (2000) The combined effect of high hydrostatic pressure, heat and bacteriocins on inactivation of foodborne pathogens in milk and orange juice. World J Microb Biot 16:387–392

Ananou S, Muñoz A, Martínez-Bueno M et al (2010) Evaluation of an enterocin AS-48 enriched bioactive powder obtained by spray drying. Food Microbiol 27:58–63

Anastasiou R, Aktypis A, Georgalaki M et al (2009) Inhibition of *Clostridium tyrobutyricum* by *Streptococcus macedonicus* ACA-DC 198 under conditions mimicking Kasseri cheese production and ripening. Int Dairy J 19:330–335

Arqués JL, Rodríguez E, Gaya P et al (2005a) Effect of combinations of high-pressure treatment and bacteriocin-producing lactic acid bacteria on the survival of *Listeria monocytogenes* in raw milk cheese. Int Dairy J 15:893–900

Arqués JL, Rodríguez E, Gaya P et al (2005b) Inactivation of *Staphylococcus aureus* in raw milk cheese by combinations of high-pressure treatments and bacteriocin-producing lactic acid bacteria. J Appl Microbiol 98:254–260

Avila M, Garde S, Gaya P et al (2006) Effect of high-pressure treatment and a bacteriocin-producing lactic culture on the proteolysis, texture, and taste of Hispanico cheese. J Dairy Sci 89:2882–2893

Benech RO, Kheadr EE, Lacroix C et al (2002a) Antibacterial activities of nisin Z encapsulated in liposomes or produced in situ by mixed culture during Cheddar cheese ripening. Appl Environ Microbiol 68:5607–5619

Benech RO, Kheadr EE, Lacroix C et al (2003) Impact of nisin producing culture and liposome-encapsulated nisin on ripening of *Lactobacillus* added-Cheddar cheese. J Dairy Sci 86:1895–1909

Benech RO, Kheadr EE, Laridi R et al (2002b) Inhibition of *Listeria innocua* in Cheddar cheese by addition of nisin Z in liposomes or in situ production by mixed culture. Appl Environ Microbiol 68:3683–3690

Benkerroum N, Oubel H, Mimoun LB (2002) Behavior of *Listeria monocytogenes* and *Staphylococcus aureus* in yogurt fermented with a bacteriocin-producing thermophilic starter. J Food Prot 65:799–805

Bizani D, Motta AS, Morrissy JAC et al (2005) Antibacterial activity of cerein 8A, a bacteriocin-like peptide produced by *Bacillus cereus*. Int Microbiol 8:125–131

Black EP, Kelly AL, Fitzgerald GF (2005) The combined effect of high pressure and nisin on inactivation of microorganisms in milk. Inn Food Sci Emerg Technol 6:286–292

Bogovič Matijašić B, Koman Rajšp M, Perko B et al (2007) Inhibition of *Clostridium tyrobutyricum* in cheese by *Lactobacillus gasseri*. Int Dairy J 17:157–166

Bouksaim M, Lacroix C, Audet P et al (2000) Effects of mixed starter composition on nisin Z production by *Lactococcus lactis* subsp. *lactis* biovar. *diacetylactis* UL 719 during production and ripening of Gouda cheese. Int J Food Microbiol 59:141–156

Buyong N, Kok J, Luchansky JB (1998) Use of a genetically enhanced, pediocin-producing starter-culture, *Lactococcus lactis* subsp. *lactis* MM217, to control *Listeria monocytogenes* in Cheddar cheese. Appl Environ Microbiol 64:4842–4845

Calderón-Miranda ML, Barbosa-Cánovas GV, Swanson BG (1999) Inactivation of Listeria innocua in skim milk by pulsed electric fields and nisin. Int J Food Microbiol 51:19–30

Čanžek Majhenič A, Bogovič Matijašić B, Rogelj I (2003) Chromosomal location of genetic determinants for bacteriocins produced by *Lactobacillus gasseri* K7. J Dairy Res 70:199–203

Cao-Hoang L, Chaine A, Grégoire L et al (2010) Potential of nisin-incorporated sodium caseinate films to control *Listeria* in artificially contaminated cheese. Food Microbiol 27:940–944

Capellas M, Mor-Mur M, Gervilla R et al (2000) Effect of high pressure combined with mild heat or nisin on inoculated bacteria and mesophiles of goats' milk fresh cheese. Food Microbiol 17:633–641

Claeys WL, Cardoen S, Daube G et al (2013) Raw or heated cow milk consumption: Review of risks and benefits. Food Control 31:251–262

Cocolin L, Innocente N, Biasutti M et al (2004) The late blowing in cheese: a new molecular approach based on PCR and DGGE to study the microbial ecology of the alteration process. Int J Food Microbiol 90:83–91

Dal Bello B, Cocolin L, Zeppa G et al (2011) Technological characterization of bacteriocin producing *Lactococcus lactis* strains employed to control *Listeria monocytogenes* in cottage cheese. Int J Food Microbiol 153:58–65

Davies EA, Bevis HE, Delves-Broughton J (1997) The use of the bacteriocin, nisin, as a preservative in ricotta-type cheeses to control the food-borne pathogen *Listeria monocytogenes*. Lett Appl Microbiol 24:343–346

Davies EA, Delves-Broughton J (1999) Nisin. In: Robinson R, Batt C, Patel P (eds) Encyclopedia of food microbiology. Academic Press, London, pp 191–198

De Buyser ML, Dufour B, Maire M et al (2001) Implication of milk and milk products in food-borne diseases in France and in different industrialized countries. Int J Food Microbiol 67:1–17

De Vuyst L, Tsakalidou E (2008) *Streptococcus macedonicus*, a multi-functional and promising species for dairy fermentations. Int Dairy J 18:476–485

Deegan LH, Cotter PD, Hill C et al (2006) Bacteriocins: biological tools for bio-preservation and shelf-life extension. Int Dairy J 16:1058–1071

Dias BE, Galer CD, Moran JW et al (2009) Cheese flavoring systems prepared with bacteriocins. US Patent 7,556,833 (Appl. No.: 10/723,257)

Ennahar S, Aoude-Werner D, Sorokine O et al (1996) Production of pediocin AcH by *Lactobacillus plantarum* WHE 92 isolated from cheese. Appl Environ Microbiol 62:4381–4387

Farías ME, Nuñez de Kairuz M, Sesma F et al (1999) Inhibition of *Listeria monocytogenes* by the bacteriocin enterocin CRL35 during goat cheese making. Milchwissenschaft 54:30–32

Fernández de Palencia P, de la Plaza M, Mohedano ML et al (2004) Enhancement of 2-methylbutanal formation in cheese by using a fluorescently tagged Lacticin 3147 producing *Lactococcus lactis* strain. Int J Food Microbiol 93:335–347

Foulquié Moreno MR, Rea MC, Cogan TM et al (2003) Applicability of a bacteriocin-producing *Enterococcus faecium* as a co-culture in Cheddar cheese manufacture. Int J Food Microbiol 81:73–84

Foulquié Moreno MR, Sarantinopoulos P, Tsakalidou E et al (2006) The role and application of enterococci in food and health. Int J Food Microbiol 106:1–24

Fox PF, McSweeney PLH, Lynch CM (1998) Significance of non-starter lactic acid bacteria in cheddar cheese. Aust J Dairy Technol 53:83–89

Franz CMAP, van Belkum MJ, Holzapfel WH et al (2007) Diversity of enterococcal bacteriocins and their grouping into a new classification scheme. FEMS Microbiol Rev 31:293–310

Gálvez A, Abriouel H, López RL et al (2007) Bacteriocin-based strategies for food biopreservation. Int J Food Microbiol 120:51–70

Gálvez A, Lopez RL, Abriouel H et al (2008) Application of bacteriocins in the control of food-borne pathogenic and spoilage bacteria. Crit Rev Biotechnol 28:125–152

Gálvez A, Valdivia E, Martínez-Bueno M et al (1990) Induction of autolysis in *Enterococcus faecalis* by peptide AS-48. J Appl Bacteriol 69:406–413

García MT, Martínez Cañamero M, Lucas R et al (2004) Inhibition of *Listeria monocytogenes* by enterocin EJ97 produced by *Enterococcus faecalis* EJ97. Int J Food Microbiol 90:161–170

García-Graells C, Masschalck B, Michiels CW (1999) Inactivation of *Escherichia coli* in milk by high-hydrostatic-pressure treatment in combination with antimicrobial peptides. J Food Prot 62:1248–1254

Garde S, Ávila M, Arias R et al (2011) Outgrowth inhibition of *Clostridium beijerinckii* spores by a bacteriocin-producing lactic culture in ovine milk cheese. Int J Food Microbiol 150:59–65

Garde S, Ávila M, Gaya P et al (2006) Proteolysis of Hispánico cheese manufactured using lacticin 481-producing *Lactococcus lactis* ssp. *lactis* INIA 639. J Dairy Sci 89:840–849

Garde S, Ávila M, Medina M et al (2005) Influence of a bacteriocin-producing lactic culture on the volatile compounds, odour and aroma of Hispánico cheese. Int Dairy J 15:1034–1043

Garde S, Tomillo J, Gaya P et al (2002) Proteolysis in Hispánico cheese manufactured using a mesophilic starter, a thermophilic starter, and bacteriocin-producing *Lactococcus lactis* subsp. *lactis* INIA 415 adjunct culture. J Agric Food Chem 50:3479–3485

Giraffa G (1995) Enterococcal bacteriocins: their potential as *anti-Listeria* factors in dairy technology. Food Microbiol 12:291–299

Giraffa G (2003) Functionality of enterococci in dairy products. Int J Food Microbiol 88(2–3):215–222

Giraffa G, Carminati D (1997) Control of *Listeria monocytogenes* in the rind of Taleggio, a surface-smear cheese, by a bacteriocin from *Enterococcus faecium* 7C5. Sci Aliment 17:383–391

Giraffa G, Carminati D, Tarelli GT (1995a) Inhibition of *Listeria innocua* in milk by bacteriocin-producing *Enterococcus faecium* 7C5. J Food Protect 58:621–623

Giraffa G, Picchioni N, Neviani E et al (1995b) Production and stability of an *Enterococcus faecium* bacteriocin during Taleggio cheesemaking and ripening. Food Microbiol 12:301–307

Gonzalez CF, Kunka BS (1987) Plasmid-associated bacteriocin production and sucrose fermentation in *Pediococcus acidilactici*. Appl Environ Microbiol 53:2534–2538

Grande MJ, Lucas R, Abriouel H et al (2006a) Inhibition of toxicogenic *Bacillus cereus* in rice-based foods by enterocin AS-48. Int J Food Microbiol 106:185–194

Grande MJ, Lucas R, Abriouel H et al (2006b) Inhibition of *Bacillus licheniformis* LMG 19409 from ropy cider by enterocin AS-48. J Appl Microbiol 101:422–428

Grattepanche F, Audet P, Lacroix C (2007) Milk fermentation by functional mixed culture producing nisin Z and exopolysaccharides in a fresh cheese model. Int Dairy J 17:123–132

Guinane CM, Cotter PD, Hill C et al (2005) Microbial solutions to microbial problems: Lactococcal bacteriocins for the control of undesirable biota in food. J Appl Microbiol 98:1316–1325

He L, Chen W (2006) Synergetic activity of nisin with cell-free supernatant of *Bacillus licheniformis* ZJU12 against food-borne bacteria. Food Res Int 39:905–909

Holo H, Faye T, Brede DA et al (2002) Bacteriocins of propionic acid bacteria. Lait 82:59–68

Iseppi R, Pilati F, Marini M et al (2008) Anti-listerial activity of a polymeric film coated with hybrid coatings doped with Enterocin 416K1 for use as bioactive food packaging. Int J Food Microbiol 123:281–287

Izquierdo E, Marchioni E, Aoude-Werner D et al (2009) Smearing of soft cheese with *Enterococcus faecium* WHE 81, a multi-bacteriocin producer, against *Listeria monocytogenes*. Food Microbiol 26:16–20

Lauková A, Czikková S (2001) Antagonistic effect of enterocin CCM 4231 from *Enterococcus faecium* on "bryndza," a traditional Slovak dairy product from sheep milk. Microbiol Res 156:31–34

Lauková A, Czikková S, Burdová O (1999) Anti-staphylococcal effect of enterocin in Sunar® and yogurt. Folia Microbiol 44(6):707–711

Lauková A, Vlaemynick G, Czikková S (2001) Effect of enterocin CCM 4231 on *Listeria monocytogenes* in Saint-Paulin cheese. Folia Microbiol 46:157–160

Le Bourhis AG, Doré J, Carlier JP et al (2007) Contribution of *C. beijerinckii* and *C. sporogenes* in association with *C. tyrobutyricum* to the butyric fermentation in Emmental type cheese. Int J Food Microbiol 113:154–163

Lee CH, Park H, Lee DS (2004) Influence of antimicrobial packaging on kinetics of spoilage microbial growth in milk and orange juice. J Food Eng 65:527–531

Liu L, O'Conner P, Cotter PD et al (2008) Controlling *Listeria monocytogenes* in Cottage cheese through heterologous production of enterocin A by *Lactococcus lactis*. J Appl Microbiol 104:1059–1066

López-Pedemonte TJ, Roig-Sagués AX, Trujillo AJ (2003) Inactivation of spores of *Bacillus cereus* in cheese by high hydrostatic pressure with the addition of nisin of lysozyme. J Dairy Sci 86:3075–3081

Lortal S, Chapot-Chartier MP (2005) Role, mechanisms and control of lactic acid bacteria lysis in cheese. Int Dairy J 15:857–871

Malheiros PS, Sant'Anna V, Barbosa MS et al (2012) Effect of liposome-encapsulated nisin and bacteriocin-like substance P34 on *Listeria monocytogenes* growth in Minas frescal cheese. Int J Food Microbiol 156(3):272–277

Martínez Viedma P, Abriouel H, Ben Omar N (2009a) Anti-staphylococcal effect of enterocin AS-48 in bakery ingredients of vegetable origin, alone and in combination with selected antimicrobials. J Food Sci 74:M384–M389

Martínez-Viedma P, Abriouel H, Ben Omar N et al (2009b) Assay of enterocin AS-48 for inhibition of foodborne pathogens in desserts. J Food Protect 72:1654–1659

Martínez-Cuesta M, Bengoechea J, Bustos I et al (2010) Control of late blowing in cheese by adding lacticin 3147-producing *Lactococcus lactis* IFPL 3593 to the starter. Int Dairy J 20:18–24

Mathot AG, Beliard E, Thuault D (2003) *Streptococcus thermophilus* 580 produces a bacteriocin potentially suitable for inhibition of *Clostridium tyrobutyricum* in hard cheese. J Dairy Sci 86:3068–3074

Mauriello G, De Luca E, La Storia A et al (2005) Antimicrobial activity of a nisin-activated plastic film for food packaging. Lett Appl Microbiol 41:464–469

McAuliffe O, Hill C, Ross RP (1999) Inhibition of *Listeria monocytogenes* in cottage cheese manufactured with a lacticin 3147-producing starter culture. J Appl Microbiol 86(2):251–256

McSweeney PLH, Fox PF (2004) Metabolism of residual lactose and of lactate and citrate. In: Fox PF, McSweeney PLH, Cogan TM et al (eds) Cheese: chemistry, physics and microbiology, vol. 1: general aspects, 3rd edn. Elsevier Academic Press, London, pp 361–371

Mills S, Serrano LM, Griffin C et al (2011) Inhibitory activity of *Lactobacillus plantarum* LMG P-26358 against *Listeria innocua* when used as an adjunct starter in the manufacture of cheese. Microb Cell Fact 10(Suppl 1):S7

Mollet B, Peel J, Pridmore D et al (2004) Bactericide compositions prepared and obtained from *Microccus varians*. US Patent 6,689,750 (Appl. No.: 08/693,353)

Morgan S, Ross RP, Hill C (1997) Increasing starter cell lysis in Cheddar cheese using a bacteriocin-producing adjunct. J Dairy Sci 8:1–10

Morgan SM, Garvin M, Ross RP et al (2001) Evaluation of a spray-dried lacticin 3147 powder for the control of *Listeria monocytogenes* and *Bacillus cereus* in a range of food systems. Lett Appl Microbiol 33:387–391

Morgan SM, O'Sullivan L, Ross RP et al (2002) The design of a three strain starter system for Cheddar cheese manufacture exploiting bacteriocin-induced starter lysis. Int Dairy J 12:985–993

Morgan SM, Ross RP, Beresford T et al (2000) Combination of hydrostatic pressure and lacticin 3147 causes increased killing of *Staphylococcus* and *Listeria*. J Appl Microbiol 88(3):414–420

Muñoz A, Ananou S, Gálvez A et al (2007) Inhibition of *Staphylococcus aureus* in dairy products by enterocin AS-48 produced in situ and ex situ: Bactericidal synergism through heat and AS-48. Int Dairy J 17:760–769

Muñoz A, Maqueda M, Gálvez A et al (2004) Biocontrol of psychrotrophic enterotoxigenic *Bacillus cereus* in a non fat hard type cheese by an enterococcal strain-producing enterocin AS-48. J Food Prot 67:1517–1521

Nes IF, Diep DB, Havarstein LS et al (1996) Biosynthesis of bacteriocins in lactic acid bacteria. Antonie van Leeuwenhoek 70:113–128

Núñez M, Rodríguez JL, García E (1997) Inhibition of *Listeria monocytogenes* by enterocin 4 during the manufacture and ripening of Manchego cheese. J Appl Microbiol 83:671–677

O'Sullivan L, Morgan SM, Ross RP et al (2002a) Elevated enzyme release from lactococcal starter cultures on exposure to the lantibiotic lacticin 481, produced by *Lactococcus lactis* DPC5552. J Dairy Sci 85:2130–2140

O'Sullivan L, O'Connor EB, Ross RP et al (2006) Evaluation of live-culture-producing lacticin 3147 as a treatment for the control of *Listeria monocytogenes* on the surface of smear-ripened cheese. J Appl Microbiol 100:135–143

O'Sullivan L, Ross RP, Hill C (2002b) Potential of bacteriocin-producing lactic acid bacteria for improvements in food safety and quality. Biochimie 84:593–604

O'Sullivan L, Ryan MP, Ross RP et al (2003) Generation of food-grade lactococcal starters which produce the lantibiotics lacticin 3147 and lacticin 481. Appl Environ Microbiol 69:3681–3685

Papagianni M, Anastasiadou S (2009) Pediocins: the bacteriocins of pediococci. Sources, production, properties and applications. Microb Cell Fact 8:3

Peláez C, Requena T (2005) Exploiting the potential of bacteria in the cheese ecosystem. Int Dairy J 15:831–844

Plockova M, Stepanek M, Demnerova K et al (1996) Effect of nisin for improvement in shelf life and quality of processed cheese. Adv Food Sci 18:78–83

Pol IE, Mastwijk HC, Slump RA et al (2001) Influence of food matrix on inactivation of *Bacillus cereus* by combinations of nisin, pulsed electric field treatment, and carvacrol. J Food Prot 64:1012–1018

Reviriego C, Fernández A, Horn N et al (2005) Production of pediocin PA-1, and coproduction of nisin A and pediocin PA-1, by wild *Lactococcus lactis* strains of dairy origin. Int Dairy J 15:45–49

Rilla N, Martínez B, Delgado T et al (2003) Inhibition of *Clostridium tyrobutyricum* in Vidiago cheese by *Lactococcus lactis* ssp. *lactis* IPLA 729, a nisin Z producer. Int J Food Microbiol 85:23–33

Roberts RF, Zottola EA, McKay LL (1992) Use of nisin-producing starter cultures suitable for Cheddar cheese manufacture. J Dairy Sci 75:2353–2363

Rodríguez E, Arques JL, Nuñez M et al (2005) Combined effect of high-pressure treatments and bacteriocin-producing lactic acid bacteria on inactivation of *Escherichia coli* O157:H7 in raw-milk cheese. Appl Environ Microbiol 71:3399–3404

Rodríguez JL, Gaya P, Medina M (1997) Bactericidal effect of enterocin 4 on *Listeria monocytogenes* in a model dairy system. J Food Prot 60:28–32

Rodríguez JM, Martinez MI, Kok J (2002) Pediocin PA-1, a wide-spectrum bacteriocin from lactic acid bacteria. Crit Rev Food Sci Nutr 42:91–121

Ross RP, Galvin M, McAuliffe O et al (1999) Developing applications for lactococcal bacteriocins. Antonie van Leeuwenhoek 76:337–346

Ryan MP, Rea MC, Hill C et al (1996) An application in Cheddar cheese manufacture for a strain of *Lactococcus lactis* producing a novel broad-spectrum bacteriocin, lacticin 3147. Appl Environ Microbiol 62(612):619

Ryan MP, Ross RP, Hill C (2001) Strategy for manipulation of cheese flora using combinations of lacticin 3147-producing and -resistant cultures. Appl Environ Microbiol 67:2699–2704

Sallami L, Kheadr EE, Fliss I et al (2004) Impact of autolytic, proteolytic and nisin-producing adjunct cultures on biochemical and textural properties of Cheddar cheese. J Dairy Sci 87:1585–1594

Scannell AG, Hill C, Ross RP et al (2000) Development of bioactive food packaging materials using immobilised bacteriocins Lacticin 3147 and Nisaplin®. Int J Food Microbiol 60:241–249

Sebti I, Delves-Broughton J, Coma V (2003) Physicochemical properties and bioactivity of nisin-containing cross-linked hydroxypropylmethylcellulose films. J Agric Food Chem 51:6468–6474

Smith K, Mittal GS, Griffiths MW (2002) Pasteurization of milk using pulsed electrical field and antimicrobials. J Food Sci 6:2304–2308

Sobrino A, Martínez Viedma P, Abriouel H et al (2009) The impact of adding antimicrobial peptides to milk inoculated with *Staphylococcus aureus* and processed by High-intensity pulsed electric field. J Dairy Sci 92:2514–2523

Sobrino-López A, Martín-Belloso O (2008) Use of nisin and other bacteriocins for preservation of dairy products. Int Dairy J 18:329–343

Sobrino-López A, Raybaudi-Massilia R, Martín-Belloso O (2006) Enhancing inactivation of *Staphylococcus aureus* in skim milk by combining high intensity pulsed electric fields and nisin. J Food Protect 69:345–353

Somkuti GA, Steinberg DH (2010) Pediocin production in milk by *Pediococcus acidilactici* in co-culture with *Streptococcus thermophilus* and *Lactobacillus delbrueckii* subsp. *bulgaricus*. J Ind Microbiol Biotechnol 37:65–69

Terebiznik MR, Jagus RJ, Cerrutti P et al (2002) Inactivation of *Escherichia coli* by a combination of nisin, pulsed electric fields, and water activity reduction by sodium chloride. J Food Prot 65:1253–1258

Thomas LV, Clarkson MR, Delves-Broughton J (2000) Nisin. In: Naidu AS (ed) Natural food antimicrobial systems. CRC-Press, Boca Raton, FL, pp 463–524

Thomas LV, Delves-Broughton J (2001) New advances in the application of the food preservative nisin. Adv Food Sci 2:11–22

Vedamuthu Ebenezer R (1995) Method of producing a yogurt product containing bacteriocin PA-1. US Patent 5,445,835 (Appl. No.: 08/192,960)

Vera Pingitore E, Todorov SD, Sesma F et al (2012) Application of bacteriocinogenic *Enterococcus mundtii* CRL35 and *Enterococcus faecium* ST88Ch in the control of *Listeria monocytogenes* in fresh Minas cheese. Food Microbiol 32(1):38–47

Weber GH, Broich WA (1986) Shelf-life extension of cultured dairy foods. C Dairy Prod J 21:19

Chapter 6
Biopreservation of Egg Products

6.1 Application of Bacteriocins

A few studies have investigated the preservation of egg and egg products by application of bacteriocins (Table 6.1). The commercial use of liquid whole egg requires processing in order to prolong its shelf-life and to inactivate foodborne pathogens. As an alternative to conventional pasteurization, an ultrapasteurization processes (i.e., heating at >60 °C for <3.5 min) was developed, which, when coupled with aseptic processing and packaging, produced liquid whole egg with a shelf life of at least 10 weeks at 4 °C. The use of effective aseptic filling and packaging systems (to prevent postpasteurization contamination) remains an essential component in the production of ultrapasteurized liquid whole egg with an extended pathogen-free shelf life. Contrary to *Salmonella*, conventional minimal egg pasteurization processes do not grant a complete inactivation of *Listeria monocytogenes*. As a matter of fact, *Listeria* species can be isolated from commercially broken raw liquid whole egg. Therefore, it was proposed to use bacteriocins for the control of *Listeria* in this food system (Schuman and Sheldon 2003). Addition of nisin to pasteurized liquid whole egg reduced the viable counts of *L. monocytogenes*, increased the product refrigerated shelf-life, and protected the liquid egg from growth of *L. monocytogenes* and *Bacillus cereus* during storage (Delves-Broughton et al. 1992; Knight et al. 1999; Schuman and Sheldon 2003). Nisin (200 IU/ml) extended the shelf life of conventionally pasteurized liquid whole egg at 6 °C by 9 to 11 days relative to nisin-free control samples (Delves-Broughton et al. 1992). The addition of nisin (1,000 IU/ml) to pH-adjusted ultrapasteurized liquid whole egg reduced *L. monocytogenes* populations by 1.6 to over 3.3 log CFU/ml and delayed (pH 7.5) or prevented (pH 6.6) the growth of the pathogen for 8–12 weeks at 4 and 10 °C (Schuman and Sheldon 2003). Both nisin and pediocin PA-1/Ach acted synergistically with heat treatments against *L. monocytogenes* (Knight et al. 1999; Muriana 1996). Nisin added at 10 mg/l significantly decreased the decimal reduction times (D-values) for *L. monocytogenes* in liquid whole egg. This effect was greater when the bacteriocin

© The Author(s) 2014

A. Gálvez et al., *Food Biopreservation*, SpringerBriefs in Food, Health, and Nutrition, DOI 10.1007/978-1-4939-2029-7_6

Table 6.1 Examples of bacteriocin applications in egg products

Bacteriocin treatment	Effect(s)	Reference(s)
Nisin	Decreased the decimal reduction times (*D*-values) for *L. monocytogenes* in liquid whole egg	Knight et al. 1999
		Schuman and Sheldon 2003
	Inactivation of *L. monocytogenes* in pasteurized liquid whole egg	
	Improves product shelf life, inhibiting post-process proliferation of *L. monocytogenes* and *B. cereus*	
Nisin and PEF	Greater inactivation of *L. innocua* in liquid egg	Calderón-Miranda et al. 1999
Nisin and HHP	Greater inactivation of *L. innocua* and *E. coli* in liquid egg	Ponce et al. 1998
Nisin in polylactic acid (PLA) coating	Inactivation of *L. monocytogenes* in liquid egg white	Jin 2010
Nisin plus allyl isothiocyanate in PLA coating	Inactivation of a three-strain *S. enterica* cocktail in liquid egg white	Jin et al. 2013

was added at least 2 h before application of heat treatments (Knight et al. 1999). In spite of the fact that nisin is not active on Gram-negative bacteria, nisin addition also increased the heat sensitivity of *Salmonella Enteritidis* PT4 in liquid whole egg and in egg white during pasteurization (Boziaris et al. 1998).

The presence of *B. cereus* in liquid egg can pose a serious hazard to the food industry, since a mild heat treatment cannot guarantee its complete inactivation. In one study, the effects of added nisin, lysozyme, or a combination of both antimicrobials on the lag phase of *B. cereus* inoculated in homogenized liquid egg samples previously heated at 60 °C for 10 min to inactivate background microflora was investigated. The combination of lysozyme and nisin delayed the average onset of growth until 10 h at 25 °C or approximately 30 h in samples stored at 16 °C (Antolinos et al. 2011).

Bacteriocins have been tested in liquid egg in combination with pulsed electric field (PEF) and high hydrostatic pressure (HHP) treatments as a way to enhance inactivation of microorganisms. Exposure of *Listeria innocua* to nisin after PEF treatment at low temperature showed an additive to synergistic effect, depending on the bacteriocin concentration and the electric field intensity and number of pulses (Calderón-Miranda et al. 1999). PEF treatments sensitized *L.innocua* to further exposure to nisin in liquid whole egg. The PEF treatment followed by the exposure of *L. innocua* to nisin increased microbial inactivation compared to the inactivation observed when the bacterium was subjected to PEF alone. The combined effect of both factors (PEF treatment and nisin addition) was either additive or synergistic, depending on the intensity of PEF treatments (Calderón-Miranda et al. 1999). Addition of nisin in combination with HHP treatment markedly reduced viable cell counts of *Escherichia coli* and *L. innocua* in liquid whole egg (Ponce et al. 1998). A reduction of almost 5 log units in *E. coli* counts and more than 6 log units for *L.*

innocua was obtained at 450 MPa and 5 mg/l nisin. For this treatment, the two microorganisms were not detectable after 1 month of storage at 4 °C. However, nisin showed no effect in preventing growth of *E. coli* in samples stored at 20 °C after pressurization. Nevertheless, counts of *L. innocua* were about 5 log cycles lower than controls after 5 days of storage at 20 °C. This could be explained by the greater reduction of viable counts obtained for this bacterium in the combined treatment, but also to a post-process protective effect of the added nisin.

Another approach for application of nisin in egg was immobilization of the bacteriocin. Liquid egg white inoculated with *L. monocytogenes* Scott A was stored in glass jars that were coated with a mixture of polylactic acid (PLA) polymer and nisin, and stored at 4 and 10 °C (Jin 2010). *Listeria* cells in control and PLA coating without nisin samples declined 1 log CFU/ml during the first 6 days at 10 °C and during 28 days at 4 °C, and then increased to 8 or 5.5 log CFU/ml. In comparison, the treatment of PLA coating with 250 mg nisin rapidly reduced the cell numbers of *Listeria* in liquid egg white to undetectable levels after 1 day, and the bacterium remained undetectable throughout the whole storage periods (48 days at 10 °C and 70 days at 4 °C). Another study reported that a PLA coating containing allyl isothiocyanate in combination with 250 mg nisin reduced the population of a three-strain *Salmonella enterica* cocktail inoculated in liquid egg white to an undetectable level after 21 days of storage (Jin et al. 2013).

References

Antolinos V, Muñoz M, Ros-Chumillas M et al (2011) Combined effect of lysozyme and nisin at different incubation temperature and mild heat treatment on the probability of time to growth of *Bacillus cereus*. Food Microbiol 28:305–310

Boziaris IS, Humpheson L, Adams MR (1998) Effect of nisin on heat injury and inactivation of *Salmonella enteritidis* PT4. Int J Food Microbiol 43:7–13

Calderón-Miranda ML, Barbosa-Cánovas GV, Swanson BG (1999) Inactivation of *Listeria innocua* in liquid whole egg by pulsed electric fields and nisin. Int J Food Microbiol 51:7–17

Delves-Broughton J, Williams GC, Wilkinson S (1992) The use of bacteriocin, nisin, as a preservative in pasteurized liquid whole egg. Lett Appl Microbiol 15:133–136

Jin T (2010) Inactivation of *Listeria monocytogenes* in skim milk and liquid egg white by antimicrobial bottle coating with polylactic acid and nisin. J Food Sci 75:M83–M88

Jin TZ, Gurtler JB, Li SQ (2013) Development of antimicrobial coatings for improving the microbiological safety and quality of shell eggs. J Food Prot 76:779–785

Knight KP, Bartlet FM, McKellar RC et al (1999) Nisin reduces the thermal resistance of *Listeria monocytogenes* Scott A in liquid whole egg. J Food Prot 62:999–1003

Muriana PM (1996) Bacteriocins for control of*Listeria* spp. in food. J Food Prot 59:S54–S63

Ponce E, Pla R, Sendra E et al (1998) Combined effect of nisin and high hydrostatic pressure on destruction of *Listeria innocua* and *Escherichia coli* in liquid whole egg. Int J Food Microbiol 43:15–19

Schuman JD, Sheldon BW (2003) Inhibition of *Listeria monocytogenes* in pH-adjusted pasteurized liquid whole egg. J Food Prot 66:999–1006

Chapter 7
Biopreservation of Seafoods

7.1 Application of Bacteriocin Preparations

Listeria monocytogenes is the main bacterial pathogen of concern in seafood products. One study found *L. monocytogenes* in ca. 30 % of smoked-fish samples, although viable counts were below 100 CFU/g (Uyttendaele et al. 2009). Another study found populations of *L. monocytogenes* greater than 10^2 CFU/g in 2.6 % of fresh fish, 5.1 % in smoked fish and 10 % in salted-fish purchased in fish farms, while 20 % of smoked fish purchased in a fish market were also contaminated (Basti et al. 2006). The bacterium was also found in raw fillets of catfish (23.5 %), trout (5.7 %), tilapia (10.3 %), and salmon (10.6 %) (Pao et al. 2008), or in 44.5 % of raw freshwater fish tested (Yücel and Balci 2010). Bacteriocin preparations have been tested singly or in combination with other hurdles to control *L. monocytogenes* in different types of seafoods (Table 7.1).

7.1.1 Raw Seafoods

Inhibition of aerobic bacteria is important to prevent seafood spoilage. The combination of nisin and Microgard™ reduced the total bacterial counts and delayed growth of *L. monocytogenes* in fresh-chilled salmon during 14 days at 6 °C, increasing the product shelf life (Zuckerman and Ben Avraham 2002; Calo-Mata et al. 2008). The observed effect was explained by the inhibitory activity of Microgard™ on Gram-negative bacteria and nisin activity on Gram-positives. Inhibition of *L. monocytogenes* in fesh salmon was also considered to be relevant as a way of preventing or reducing the levels of this bacterium in processed products such as cold-smoked salmon. In another study, nisin (200 IU/g) added on fresh gilthead seabream fillets packed under modified atmosphere was the most effective treatment resulting in significant shelf life extension of fillets (48 days compared to 10 days

© The Author(s) 2014   75
A. Gálvez et al., *Food Biopreservation*, SpringerBriefs in Food, Health,
and Nutrition, DOI 10.1007/978-1-4939-2029-7_7

Table 7.1 Example applications of bacteriocin and bacteriocin-producing LAB in seafoods

Bacteriocin treatment	Effect(s)	Reference(s)
Nisin- radio frequency heating at 65 °C	Complete inactivation of *L. innocua* in sturgeon caviar or ikura	Al-Holy et al. 2004
Nisaplin	Inactivation of *L. monocytogenes* in red-pepper seasoned cod roe	Hara et al. 2009
Nisin-coated plastic films	Inactivation of *L. monocytogenes* in CSS during refrigeration storage	Neetoo et al. 2008a
	Inhibition of background spoilage microbiota	
Broad-spectrum bacteriocin preparation from *L. lactis* PSY2	Reduction of total viable counts on the surface of reef cod fillets	Sarika et al. 2012
Enterocin AS-48	Inhibition of biogenic amine forming LAB on sardine fillets	Ananou et al. 2014
C. divergens V41 or its culture supernatant	Inhibitory effect on *L. innocua* 2030c growth cold-smoked salmon-trout	Vaz-Velho et al. 2005
C. divergens M35, or divergicin M35	Suggested as bio-ingredient for application to the inactivation of *L. monocytogenes* in ready-to-eat seafood	Tahiri et al. 2009b
E. mundtii	Inhibition of of *L. monocytogenes* in CSS	Bigwood et al. 2012
L. curvatus CWBI-B28 culture, spraying with bacteriocin, packaging in bacteriocin-coated plastic film, cell-adsorbed bacteriocin	Variable inactivation of *L. monocytogenes* in CSS. Best results were reported for bacteriocin adsorbed on heat-inactivated cells	Ghalfi et al. 2006
L. curvatus ET30	Reduction of *L. innocua* counts on salmon fillets before and after cold-smoking and during vacuum pack storage	Tomé et al. 2008
Leuconostoc spp., *L. fuchuensis, C. alterfunditum*	Broad-spectrum bio-protective cultures	Matamoros et al. 2009
Bifidobacterium-thymol combination	Extended shelf life of fresh plaice fillets; inhibition of fresh packaged fish spoilers	Altieri et al. 2005
S. xylosus	Use as protective culture to decrease biogenic amine formation in salted and fermented anchovy	Mah and Hwang 2009
L. lactis (nisin-producer)	Use as starter culture to improve Senegalese guedj fish fermentation	Diop et al. 2009

for the control at 0 °C; Tsironi and Taoukis 2010). Also, when nisin was tested in combination with other antimicrobials (such as the lactoperoxidase system or with headspace CO_2 levels and EDTA), increased inactivation or growth delay of spoilage microbiota was observed in sardines and in fish muscle extract.

In one recent study, enterocin AS-48 was applied on sardine fillets immersion in a bacteriocin solution (250 mg/l) for 1 min (Ananou et al. 2014). The treated samples were refrigeration stored under normal, vacuum, or modified atmosphere packaging. The application of enterocin AS-48 did not reduce the mesophilic, psychrotrophic, or Gram-negative bacteria viable cell counts under any of the storage conditions tested. AS-48 did cause significant reductions in viable staphylococci counts, especially under vacuum packaging. Storage of samples treated with enterocin AS-48 under modified atmosphere or under vacuum packaging allowed reductions (significant at some storage times) in histamine- and tyramine-forming LAB. The most interesting results of this study are those concerning the observed decrease (by several fold) in the levels of the biogenic amines cadaverine, putrescine, tyramine, and histamine determined after treatment with AS-48.

Antibacterial activity with broad inhibitory spectrum (*Arthrobacter* sp., *Acinetobacter* sp., *Bacillus subtilis*, *Escherichia coli*, *L. monocytogenes*, *Pseudomonas aeruginosa* and *Staphylococcus aureus*) was detected in *Lactococcus lactis* PSY2 isolated from the body surface of marine perch (Sarika et al. 2012). Surface-application of a bacteriocin preparation derived from this strain on fillets of reef cod reduced bacterial growth during storage at 4 °C. The total viable count revealed that PSY2-treated fish samples remained within the maximum limit of acceptability (10^7 counts/g, according to International Commission of Microbiological Standards for Foods 1986) until day 21, while the untreated controls became unacceptable before the 14th day of storage (Sarika et al. 2012). The maximum inhibitory effect of bacteriocin PSY2 was observed against *Staphylococcus* sp. and *Pseudomonadaceae* which were reduced by 1.8 and 3.37 log units in the PSY2 treated fillets compared to the control. Acceptability in terms of sensory attributes was significantly higher in the bacteriocin-treated samples.

Data on bacterial food poisoning associated to consumption of seafoods are comparatively more scarce compared to data on incidence of *L.monocytogenes*. However, a recent study reported that enterotoxigenic *Bacillus cereus* can grow on the surface of fresh salmon at abusive temperatures, with generation times of 169.7, 53.5, and 45.6 min were at 12, 16, and 20 °C (Labbé and Rahmati 2012). Nonhemolytic enterotoxin was detected on salmon after 20 h at 20 °C and after 26 h at 16 °C when levels of *B. cereus* were in excess of 10^8 CFU/g, indicating that fresh salmon can serve as an excellent substrate for enterotoxigenic *B. cereus* and that this organism can reach levels associated with foodborne illness following moderate temperature abuse (Labbé and Rahmati 2012). Nisin, at concentrations of 1 and 15 mg/g of salmon, reduced the levels of *B. cereus* by 2.5- and 25-fold, respectively after 48 h incubation at 16 °C, although the effect of added nisin on enterotoxin production was not reported (Labbé and Rahmati 2012).

7.1.2 Ready-to-Eat Seafoods

Minimally processed refrigerated ready-to-eat seafoods can pose health risk to susceptible individuals due to contamination by *L. monocytogenes*. Proliferation of *L. monocytogenes* in slightly processed products which are consumed without further cooking (such as cold-smoked seafood products) is a matter of concern, and therefore extensive work has been carried out on application of bacteriocins in this field (O'Sullivan et al. 2002; Chen and Hoover 2003; Drider et al. 2006; Calo-Mata et al. 2008; Galvez et al. 2008). Several bacteriocins (such as nisin, psiocins, divercin or sakacin P) have been tested, either by addition of bacteriocin preparations (either by immersion, spraying, or mixing with food matrix) bacteriocin injection, or immobilisation on plastic films or coatings. In vacuum-packaged CSS stored at 10 °C for 3 weeks, purified sakacin P added at 12 ng/g or 3.5 µg/g partially or completely inhibited growth of *L. monocytogenes* (Aasen et al. 2003). It was also observed that bacteriocin titres in salmon tissue decreased during storage, which was attributed to proteolytic degradation of the bacteriocin.

Duffes et al. (1999a) tested the effect of adding semipurified divercin V41 from *Carnobacterium divergens* V41 (isolated from trout intestine) and piscicolins from *Carnobacterium piscicola* V1 on *L. monocytogenes* inoculated in CSS stored at 4 or 8 °C. Crude extracts of piscicocins were bactericidal at 4 °C and 8 °C, while divercin V41 inhibited (4 °C) or delayed (8 °C) growth. In CSS stored at 10 °C, purified divergicin M35 (50 µg/g) as well as concentrated culture supernatants reduced viable counts of *L. monocytogenes* by 1 log CFU/g at the beginning of storage and inhibited or retarded growth for up o 21 days (Tahiri et al. 2009a, b). Total lactic acid bacteria counts were not affected by bacteriocin addition. Vaz-Velho et al. (2005) applied a different strategy, based on immersion of salmon-trout fillets for 30 s in diluted *C. divergens* V41 supernatant, before the cold-smoking process. Two trials were carried out, one in summer, where temperature during the smoking process reached 33 °C, and one in winter with a lower temperature. In the first trial, the bacteriocin treatment achieved a maximum 3-log cycles reduction of *Listeria innocua* viable counts at week 1 of storage, followed by regrowth of the bacterium. In the second trial, a stronger listericidal effect was obtained. No cells of *L. innocua* were found after smoking or at the end of the storage period (Vaz-Velho et al. 2005). These results underline the influence of food processing conditions on the efficacy of bacteriocins, particularly those conditions affecting growth of the target bacterium.

In vacuum-packaged CSS, growth of *L. monocytogenes* could be prevented by a combination of carbon dioxide, nisin, NaCl, and low temperature (Nilsson et al. 1997). Preservation of vacuum packed CSS stored at 5 °C with nisin (500 or 1,000 IU/g) initially reduced the cell numbers of *L. monocytogenes* but did not further prevent growth of survivors (Nilsson et al. 1997), a behavior typically observed when bacteriocins are used at sub-inhibitory concentrations. For that reason, nisin was tested in combination with other antimicrobials, including modified atmosphere packaging, in order to improve control of this pathogen. In experiments carried out in a culture broth, the antilisterial effect of nisin was improved in the presence of 100 % CO_2 and increasing NaCl concentrations (0.5–5.0 % w/v). In CSS packed under MAP (70 %/30 % CO_2/N_2), addition of 500 and 1,000 IU nisin/g inhibited growth of *L. monocytogenes* considerably, with lag phases of 8 and 20 days, respectively, and the levels of *L. monocytogenes* remained below 10 CFU/g during 27 days of storage at both concentrations of nisin (Nilsson et al. 1997).

On smoked salmon slices inoculated with *L. monocytogenes* and surface-treated with nisin (400 or 1,250 IU/g) or ALTA™ 2341 (1 %), both antimicrobials reduced the growth of listeria to some extent (Szabo and Cahill 1999). When the treated smoked salmon was packaged in 100 % CO_2, counts of *L. monocytogenes* were reduced below detectable levels (2 logs) in both cases during 21 days of storage at 4 °C. Addition of these bacteriocins also showed a potential preservation and safety advantage if product was exposed to short-term temperature abuse (Szabo and Cahill 1999).

Chemical preservatives potassium sorbate, sodium lactate and sodium diacetate have strong inhibitory effects on growth of *L. monocytogenes* in CSS. For smoked salmon fillets, the most effective treatment was 2.4 % sodium lactate/0.125 % sodium diatetate, which was able to inhibit the growth of *L. monocytogenes* in smoked salmon fillets for 4 weeks of storage at 4 °C. Nisin showed greatest inhibitory effects in combination with potassium sorbate (0.00125 % nisin/0.15 % potassium sorbate), being able to inhibit the growth of *L. monocytogenes* to levels below 4 log CFU/g for 3 weeks (Neetoo et al. 2008b). Addition of nisin or sodium lactate also inhibited the growth of *L. monocytogenes* in cold-smoked rainbow trout, but the combination of the two compounds was more effective (Nykanen et al. 2000). Nisin, sodium lactate or their combination were injected into rainbow trout at an industrial scale before the smoking process, or injected into the finished smoked product. Best results were obtained when the combination of nisin and sodium lactate (120–180 IU nisin/g + 18 g lactate/ kg) were injected into the smoked fish, decreasing the count of *L. monocytogenes* from 3.26 to 1.8 log CFU /g over 16 days of storage at 8 °C. In the fish injected before smoking, the combination of 3.6 % sodium lactate and 240–360 IU/g nisin or 1.8 % sodium lactate and 120–180 IU/g nisin inhibited growth of *L. monocytogenes* (to almost constant levels of 4.7–4.9 log CFU/g) for 29 days at 3 °C in the vacuum-packed cold smoked samples (Nykanen et al. 2000). In both cases (before or after smoking), application of the combined antimicrobial treatments did also reduce mesophilic aerobic counts in the cold-smoked product. The treatments did not affect the sensory characteristics.

The efficacy of bacteriocins in seafoods can improve with immobilisation in packaging materials. Packaging CSS in plastic film coated with bacteriocin from *Lactobacillus curvatus* CWBI-B28 caused *L. monocytogenes* inactivation late during refrigeration storage (Ghalfi et al. 2006). However, best results (complete inactivation of *L. monocytogenes* during storage for 22 days) were reported for CSS treated with bacteriocin adsorbed to its heat-inactivated producer cells. In CSS vacuum-packed inside nisin-coated plastic films, nisin (2,000 IU/cm^2) inhibited the proliferation of a cocktail of *L. monocytogenes* strains (Neetoo et al. 2008a). Viable counts were 3.9 log CFU/cm^2 lower compared with controls for samples inoculated with 5×10^2 CFU/cm^2 of *L. monocytogenes* after 56 days of storage at 4 °C or 49 days at 10 °C. In addition, nisin inhibited the proliferation of background microbiota (aerobic, anaerobic, and LAB counts) on smoked salmon at both storage temperatures although the bacteriostatic effect was much more pronounced at 4 °C (Neetoo et al. 2008a). Chitosan dosed with 500 IU/cm^2 nisin slowed down growth of *L. monocytogenes* on CSS for 10 days at room temperature by approximately 1 log unit (Ye et al. 2008). Chitosan-coated plastic films dosed with sodium lactate (2.3 mg/cm^2) in combination with nisin (500 IU/cm²) inhibited the proliferation of *L. monocytogenes* in the vacuum-packaged CSS during storage at 4 °C for up to 6 weeks (Ye et al. 2008). The addition of nisin allowed a reduction in the concentration of sodium lactate from 4.5 mg/cm² (when tested singly) to 2.3 mg/cm² in the combination while achieving a similar antilisterial effect at least during the first 4 weeks of storage.

In another study, nisin and lysozyme (from hen egg and from oysters) were tested on CSS stored at 4 °C (Datta et al. 2008). The combinations of the two antimicrobials were applied directly or in calcium alginate coating. The effectiveness of oyster lysozyme or hen egg white lysozyme was enhanced when added incorporated onto calcium alginate coatings. After 35 days at 4 °C the growth of *L. monocytogenes* and *Salmonella* Anatum was suppressed in the range of 2.2–2.8 log CFU/g with nisin-lysozyme (from oysters as well as from hen egg) calcium alginate coatings compared to the control nontreated samples, with no significant differences from the source of lysozyme. The study concluded that calcium alginate coatings containing oyster lysozyme and nisin could be used to control the growth of *L. monocytogenes* and *Salmonella* Anatum on the surface of ready-to-eat smoked salmon at refrigerated temperatures.

Bacteriocins have been tested also in other ready-to-eat seafood products with the purpose of enhancing shelf life or reducing the risk of *L. monocytogenes*. In smoked salmon paté, results reported for nisin were not satisfactory compared to organic acid salts of potassium sorbate and sodium lactate or sodium diacetate (Neetoo et al. 2008b). Pâté samples supplemented with organic acid salt treatments had lower counts by the end of 3 weeks compared to those incorporating nisin or nisin with organic acid salts at a lower concentration. It was suggested that the ingredients in pâté provided nisin protection for *L. monocytogenes* (Neetoo et al. 2008b). In fish spreads made from hake flesh, enterocins 1071A and 1071B inhibited the growth of aerobic mesophilic bacteria during cold storage (Dicks et al. 2006). An enterocin concentrate obtained by ammonium sulphate precipitation was added to the fish spreads at 1.0 % (w/w), equivalent to 1.2×10^5 AU/g fish spread. The number of microbial cells recorded in fish spread preserved with enterocins was 8×10^6 CFU/g after 21 days of cold storage (4 °C), compared to 1×10^8 CFU/g in fish spread that had not been preserved. Enterocins 1071A and 1071B did preserve the fish spread, but to a lesser extent than a combination of sodium benzoate and potassium sorbate.

Cooked shelled crabmeat is prone to cross contamination with raw product by personnel and from raw crab in the processing environment. Washing crabmeat with antimicrobials (PerLac 1911, Microgard™, Alta™ 2341, nisin, or *Enterococcus faecium* 1083 culture supernatant containing the bacteriocin-like substance (BLIS) enterocin 1083 reduced the viable counts of *L. monocytogenes* during storage at 4 °C only in the samples treated with 20,000 AU of Alta™ 2341, nisin, or enterocin 1083 (Degnan et al. 1994). However, best results were reported for sodium diacetate or trisodium phosphate. While trisodium phosphate considerably increased the pH of crabmeat to unacceptable levels, sodium diacetate did not produce adverse effects and reduced the levels of *L. monocytogenes* by 2.6 log units/g within 6 day.

Shucked lobster meat is prepared from lobsters cooked by immersion in boiling water, cooled, and shucked by hand. The meat portions are packed with brine and

eaten without further cooking. The risks of bacterial contamination during processing and the zero tolerance level imposed by administrations for *L. monocytogenes*
in ready-to-eat seafood products strengthen the needs to apply additional preservation methods. In one study, the combined effect of nisin and moderate heat on
L. monocytogenes in cans of cold-pack lobster was investigated (Budu-Amoako
et al. 1999). Heat processing lobster meat in the presence of nisin (25 mg/kg of can
content) at 60 °C internal temperature for 5 min achieved 3- to 5-log reductions in
L. monocytogenes viable cell counts, whereas nisin or heat alone achieved 1- to
3-log reductions. The effect of the combined treatment was considered to be satisfactory, since reported *L. monocytogenes* levels in the commercial product never
exceeded 10^2 CFU/g. An additional advantage was that a reduced heat process in
combination with nisin allowed a considerable reduction in drained weight loss
compared to the standard heating process used by industry.

   In brined shrimp, addition of carnocin UI49 (from *C. piscicola* UI49) did not
extend the shelf life, while crude bavaricin A (a cell-free supernatant of *Lactobacillus
bavaricus* MI 401) resulted in a shelf life of 16 days, while a nisin Z preparation
extended the shelf life for 31 days (Einarsson and Lauzon 1995). The addition of
nisin Z and bavaricin A preparations extended the product shelf-life, although the
efficacy of bacteriocin treatments was much more limited compared to brined
shrimp stored in a benzoate-sorbate solution (Einarsson and Lauzon 1995). It was
observed that the dominant microbiota towards the end of treatments was domi-
nated by Gram-positive bacteria in samples treated with carnocin UI49 and bavaricin A as well as in the untreated controls, while in the nisin Z treatment a
Gram-negative microbiota was more pronounced. In another study, it was shown
that dipping in organic acids solutions followed by vacuum packaging and chilled
storage can help reduce *L. monocytogenes* and native microbiota, but not *Salmonella*,
on fresh shrimps (Wan Norhana et al. 2012). In that study, beheaded and peeled
fresh shrimps dipped in solutions containing nisin (500 IU/ml), EDTA (0.02 M),
potassium sorbate (PS) (3 %, w/v), sodium benzoate (SB) (3 %, w/v) or sodium
diacetate (SD) (3 %, w/v) alone or in combination were vacuum packaged and
stored at 4 °C for 7 days. Nisin-EDTA-potassium sorbate and nisin-EDTA-sodium
diacetate significantly reduced *L. monocytogenes* numbers, but none of treatments
reduced *Salmonella* counts on shrimps throughout storage. Overall, the applied
treatments improved the microbiological quality of shrimps. For example, on day 7
for storage, numbers of aerobic bacteria, psychrotrophic bacteria and *Pseudomonas*
on combined nisin-EDTA-salt of organic acids treated shrimps were significantly
lower by 4.40–4.60, 3.50–4.01, and 3.84–3.99 log CFU/g respectively, as compared
to the control (Wan Norhana et al. 2012).

   Caviar is heat labile, and conventional pasteurization processes affect its texture,
color, and flavor negatively. Refrigerated storage is currently the only available
means to preserve and extend the shelf life of caviar as a ready-to-eat product.
Chum salmon caviar (ikura) and sturgeon caviar were treated by immersion in
500 IU/ml nisin solution and heat processed (an 8-D process without nisin or a 4-D
process with 500 IU/ml nisin) in a radio frequency (RF; 27 MHz) heating method at
60, 63, and 65 °C (Al-Holy et al. 2004). The combination of RF heating and nisin

acted synergistically to inactivate *L. innocua* cells and total mesophilic microorganisms. No surviving *L. innocua* were recovered in the caviars after application of the nisin-RF combined treatments at 65 °C. The come-up times in the RF-heated product were significantly lower compared with the water bath–heated caviar at all treatment temperatures. The visual quality of the caviar products treated by RF with or without nisin was comparable to the untreated control. The effect of nisin, chemical antimicrobials or moderate heat (singly or in combination) on inactivation of *L. monocytogenes* in sturgeon caviar was further investigated. Treating caviar with 500 or 750 IU/ml nisin initially reduced *L. monocytogenes* by 2–2.5 log units (Al-Holy et al. 2005). Nisin in combination with 2 % lactic acid plus 134 ppm chlorous acid reduced viable counts of *L. monocytogenes* below detectable levels at several points during storage at 4 °C, and it also reduced total mesophilic viable counts. However, best results were obtained for the combinations of nisin and mild heat (60 °C for 3 min). Mild heating in combination with nisin synergistically reduced viable counts of *L. monocytogenes* and total mesophiles. No *L. monocytogenes* cells were recovered from caviar treated with heat and nisin (750 IU/ml) after a storage period of 28 days at 4 °C.

L. monocytogenes can be highly prevalent in minced tuna and fish (salmon and cod) roe, where it can multiply more rapidly under temperature-abuse conditions (Takahashi et al. 2011). Such seafood products are among the most popular sushi ingredients for consumers of all age groups. Since the complete elimination of *L. monocytogenes* from the processing environments in which minced tuna and fish roe products are prepared considered to be very difficult, it is necessary to apply other preservation methods that, at the same time, do not negatively affect the taste of these seafood products (Takahashi et al. 2011). So in one study, nisin in the form of Nisaplin and other antimicrobials (lysozyme, e-polylysine, and chitosan) were tested for inhibition of *L. monocytogenes* in the seafoods stored at 10 °C (Takahashi et al. 2011). Nisaplin effectively inhibited *L. monocytogenes* growth in minced tuna at 500 ppm and in salmon roe at 250 ppm within their standard shelf lives. Consequently, 500 ppm of Nisaplin, (which is the legal standard for cheese and meat products in many countries), was considered an appropriate and safe concentration for seafood.

Karashi-mentaiko is red-pepper seasoned cod roe. However, *L. monocytogenes* has been isolated from Karashi-mentaiko, and since there is no heat treatment in the manufacturing process of Karashi-mentaiko, the control of bacteria is very important (Hara et al. 2009). Nisin can effectively inhibit growth of *L. monocytogenes* in Karashi-mentaiko (Hiwaki et al. 2007). In tests carried out independently on eight different strains, the number of *L. monocytogenes* in Karashi-mentaiko stored at 4 °C was decreased by Nisaplin added at 60 and 600 µg/g (Hara et al. 2009). In the samples containing 60 µg/g Nisaplin, most of the isolates were undetected (except for two strains) through the whole storage period (28 days), while at 600 µg/g Nisaplin none of the strains were detected. When samples were stored at 15 °C, both concentrations of Nisaplin (60 and 600 µg/g) brought all strains undetectable during the whole storage period. Interestingly, the MICs for Nisaplin obtained in Karashi-mentaiko were lower compared to BHI broth, suggesting that ingredients of Karashi-mentaiko, storage temperature and a_w influenced the efficacy of Nisaplin.

7.2 Application of Bacteriocin-Producing Strains

Bacteriocin production by fish-acclimatized bacterial species is of great interest for inhibition of pathogenic microorganisms in seafood products (Table 7.1). Antagonistic bacterial strains (such as those isolated from cold smoked seafood products) could be applied for the competitive exclusion of *L. monocytogenes* in the processed food products. Many LAB strains are able to grow at refrigeration temperatures. They tolerate modified atmosphere packaging, low pH, high salt concentrations, and the presence of additives such as lactic acid, ethanol, or acetic acid. The selected antagonistic strains should meet several criteria: (1) to be able to grow on the fish product during cold storage and produce antimicrobials to inactivate *L. monocytogenes*, or at least inhibit growth of the pathogen; (2) do not cause adverse effects on the food product (such as off flavours, colour changes); (3) do not have adverse effects on health (e.g., production of biogenic amines) or carry antibiotic resistance or virulence traits. Inoculated strains could have probiotic properties, but this approach for administration of probiotics through seafood products has not been exploited yet.

Since LAB comprise the dominant microbiota in CSS (González-Rodríguez et al. 2002; Cardinal et al. 2004), research has focused on selection of antagonistic LAB strains from the processed products. *L. monocytogenes* can be inhibited by carnobacteria cultures that do not produce bacteriocins, partly due to glucose depletion (Nilsson et al. 2005). However, LAB strains producing bacteriocins (mainly *Carnobacterium* and *Lactobacillus* species) may be superior for biopreservation compared to non-bacteriocinogenic strains. Examples of trials carried out with antagonistic bacteria in CS foods (such as CSS, cold-smoked salmon-trout, or cold-smoked surubim) included antagonistic strains of *C. piscicola*, *C. divergens*, *Lactobacillus sakei*, *Lactobacillus casei*, *Lactobacillus curvatus*, *Lactobacillus delbrueckii*, *Lactobacillus plantarum*, *Pediococcus acidilactici* or *E. faecium* (Leisner et al. 2007; Calo-Mata et al. 2008; Galvez et al. 2008; Rihakova et al. 2009; Tomé et al. 2008; Tahiri et al. 2009b).

C. piscicola is often isolated as the naturally dominant LAB species on CSS (Paludan-Müller et al. 1998). Therefore, CSS may be a good source for isolation of bacteriocin-producing LAB. The strains *C. piscicola* A9b (producer of carnobacteriocin B2) and *C. piscicola* CS526 (producer of piscicolin CS526) showed anti-Listeria activity in salmon juice and in CSS, respectively (Nilsson et al. 2004; Yamazaki et al. 2003). Also, the strains *C. divergens* V41 and *C. piscicola* V1 from processed seafoods were reported to be highly effective against *L. monocytogenes* in co-culture experiments carried out in a simulated cold smoked fish system at 4 °C (Duffes et al. 1999b). In cold-smoked surubim (a native Brazilian freshwater fish), inhibition of *L. monocytogenes* by the bacteriocinogenic strain *C. piscicola* C2 isolated from vacuum-packed cold-smoked surubim and by other *C. piscicola* strains isolated from CSS was reported (Alves et al. 2005). Strong inhibition was detected both in fish peptone model systems and in cold-smoked fish juices. Although the carnobacteria grew poorly on cold-smoked surubim at 10 °C, the strains were able to reduce maximum *Listeria* counts by 1–3 log units in an artificially inoculated surubim. In another study, when an "alecrim pimenta" extract was tested in combination with strains of *Carnobacterium maltaromaticum* (bacteriocinogenic or not), variable effects were observed on *L. monocytogenes* (dos Reis et al. 2011), ranging from strong inhibition to only transient inhibition, depending on the substrate (surubin broth, or surubin homogenate). It was concluded that the use of alecrim pimento extract and cultures of carnobacteria have potential to inhibit *L. monocytogenes* in fish systems and the applications should be carefully studied, considering the influence of food matrix.

Experimental work carried out with bacteriocin-producing lactobacilli indicates that these bacteria can inhibit *L. monocytogenes* in seafood products. Cultures of *L. sakei* L6790 (producer of sakacin P) inoculated on CSS only had a bacteriostatic effect on *L. monocytogenes*, similarly to an isogenic (bac⁻) *L. sakei* strain. However, application of the bacteriocinogenic culture in combination with a sub-lethal concentration of purified sakacin P resulted in a partial inactivation of *L. monocytogenes* population (Katla et al. 2001). In another study, the bacteriocinogenic strain *L. curvatus* CWBI-B28 was tested against *L. monocytogenes* in CSS during storage at 4 °C by using different approaches: producing bacteriocin *in situ*, spraying with partially purified bacteriocin, packaging in bacteriocin-coated plastic film, and cell-adsorbed bacteriocin (a suspension of producer cells on which maximum bacteriocin has been immobilized by pH adjustments) (Ghalfi et al. 2006). In spite of the fact that all different treatments achieved some inactivation of *L. monocytogenes* in CSS, the cell-adsorbed bacteriocin provided best results, with complete inactivation of listeria for up to 20 days (Ghalfi et al. 2006). Vescovo et al. (2006) evaluated the biopreservative potential of three antimicrobial-producing LAB strains (*L. casei*, *L. plantarum* and *C. piscicola*) on refrigerated CSS stored under vacuum. All three strains were able to inhibit growth of *L. innocua* and none affected negatively the sensory quality of the product. The combination of *L. casei-L. plantarum* was the most effective in inhibition of *Listeria*, while a *L. casei–C. piscicola* association was less effective than *C. piscicola* alone (Vescovo et al. 2006). In another study, bacteriocin-producing LAB (*E. faecium* ET05, *L. curvatus* ET06, *L. curvatus* ET30, *L. delbrueckii* ET32 and *P. acidilactici* ET34), selected by their capacity for growth and producing inhibition in vitro under conditions simulating cold-smoked fish (at high salt-on-water content, low temperature and anaerobic atmosphere) were inoculated onto salmon fillets in co-culture with *L. innocua* 2030c, and cold-smoked processed. The finished product was then packed under vacuum and stored at 5 °C (Tomé et al. 2008). *L. curvatus* ET30 and *L. delbrueckii* ET32 showed a good biopreservation potential for CSS, while *L. curvatus* ET06 and *P. acidilactici* ET34 showed a bacteriostatic mode of action against the target bacteria in vitro as well as when inoculated into the salmon fillets. Comparatively, strain *E. faecium* ET05 showed the best results in controlling *L. innocua* growth in vacuum-packaged CSS processed under the salting/drying/smoking.

Bioprotective cultures can also be applied during the dry-salting process. For example, commercial preparations of *P. acidilactici* (Fargo-35, Laboratorios Amerex S.A., Madrid, Spain), *L. curvatus* (InhiList-2, Innaves S.A., Pontevedra,

Spain), and a mixture of lactic cultures and species extract (Biamex-01, Laboratorios Amerex S.A., Madrid, Spain) were tested on fresh salmon in combination with a dry salt-sugar mix and stored at 8 °C (Montiel et al. 2013). All antimicrobials tested were effective in inhibiting growth of *L. monocytogenes*. After 7 days of storage, the biopreservative based on *P. acidilactici* strongly inhibited growth of the pathogen, with counts 3.6 and 1.5 log CFU/g lower than in the control and salt-sugar mix-treated samples, respectively (Montiel et al. 2013).

Research in enterococci from seafoods has gained interest for biocontrol of *L. monocytogenes* in the processed products, and bacteriocin-producing strains have been isolated from seafoods such as *E. faecium* and *Enterococcus mundtii* strains producers of unknown bacteriocins (Campos et al. 2006; Hosseini et al. 2009; Valenzuela et al. 2010), or *E. faecium* strains producers of enterocin P (Arlindo et al. 2006) or enterocin B (Pinto et al. 2009). *Enterococcus* isolates from sea bass and sea bream showed broad antimicrobial activities (against *Carnobacterium* sp., *Bacillus* sp., *L. monocytogenes*, *Aeromonas salmonicida*, *Aeromonas hydrophila* and *Vibrio anguillarum*) and carried enterocin genes (including enterocins A, B, L50 and P), strengthening the potential applications of these LAB strains to the biopreservation of minimally-processed seafood products (Chahad et al. 2012).

Also, enterococci from other sources have been suggested for application in seafoods. An *E. mundtii* strain isolated from soil with strong anti-*Listeria* activity was tested in vacuum-packed CSS stored at 5 °C (Bigwood et al. 2012). This strain inhibited the growth of *L. monocytogenes* on the CSS during its 4 week shelf life. When *L. monocytogenes* (ca. 3.4 log CFU/cm^2) was co-inoculated with *E. mundtii* (7 log CFU/cm^2), growth of the pathogen was reduced compared to the control samples with an approximate 3 log CFU/cm^2 difference in concentration after 4 weeks incubation (Bigwood et al. 2012). The inhibitory effect was found to be dependent on the initial inoculum of enterococci. The *E. mundtii* isolate was able to grow at 5 °C in culture medium, but not on the CSS, and for that reason a high inoculum of enterococci was required in order to achieve a strong inhibition of the listeriae. The study concluded that *E. mundtii* could control the growth of *L. monocytogenes* at low temperatures, indicating a potential application in controlling this pathogen in chilled foods.

While most studies have focused on inhibition of *L. monocytogenes* in seafood products, other pathogenic bacteria (such as *Vibrio parahaemolyticus*, *Vibrio vulnificus* and *Vibrio cholerae*, *Clostridium botulinum*, histamine-producing bacteria, and post-contaminating bacteria, such as *S. aureus* or *Salmonella* sp.) or spoilage bacteria (such as *Shewanella putrefaciens*, *Photobacterium phosphoreum*, *Aeromonas* spp. and *Pseudomonas* spp.) are still a matter of concern (Gram and Dalgaard 2002; Calo-Mata et al. 2008). Therefore, there is a growing interest to extend the spectrum of inhibition of bacteriocins in combination with other hurdles and also on isolation of LAB strains with broader spectrum of inhibitory activity. One study reported that treatment of whole shrimp with potassium sorbate in combination with *Bifidobacterium breve* cells extended the product microbiological shelf life (Al-Dagal and Bazaraa 1999). Also, treatment of plaice fillets with a preparation of *Bifidobacterium bifidum* cells and thymol combined with low storage temperature and anoxia/hypoxia, showed a great efficacy against the main fresh packaged fish spoilage species (Altieri et al. 2005). Selected strains of *Leuconostoc gelidum*, *Lactococcus piscium*, *Lactobacillus fuchuensis*, and *Carnobacterium alterfunditum* (psychrotrophs, lacking antibiotic resistance traits and unable to produce histamine or tyramine) have been investigated as broad-spectrum bio-protective cultures in fish preservation (Matamoros et al. 2009).

7.3 Bacteriocin Cultures in Fermented Fish

Fermented fish products are very popular in the Asiatic and Pacific regions, but their microbiological aspects are not known so well as other fermented foods. Salted fermented foods contain abundant amino acids, which can generate relatively large amounts of biogenic amines (Mah et al. 2003). Inoculation with *Staphylococcus xylosus* producer of a BLIS has been proposed as a protective culture to decrease biogenic amine formation in salted and fermented anchovy (Mah and Hwang 2009).

L. lactis subsp. *lactis* strain CWBI B1410 (which produces various antibacterial compounds including organic acids and nisin) was tested to improve the traditional Senegalese fish fermentation into guedj (Diop et al. 2009). The inoculated starter (in combination with glucose addition) released nisin onto the fish fermentate, produced a faster acidification and reduced the counts of enteric bacteria in the fermented fish. The authors proposed a new fish fermentation strategy based on inoculation with this strain as starter, combined with salting and drying to enhance the safety of guedj.

References

Aasen IM, Markussen S, Møretrø T et al (2003) Interactions of the bacteriocins sakacin P and nisin with food constituents. Int J Food Microbiol 87:35–43

Al-Dagal MM, Bazaraa WA (1999) Extension of shelf life of whole and peeled shrimp with organic acid salts and bifidobacteria. J Food Prot 62:51–56

Al-Holy M, Lin M, Rasco B (2005) Destruction of *Listeria monocytogenes* in sturgeon (*Acipenser transmontanus*) caviar by a combination of nisin with chemical antimicrobials or moderate heat. J Food Prot 68:512–520

Al-Holy M, Ruiter J, Lin M et al (2004) Inactivation of *Listeria innocua* in nisin-treated salmon (*Oncorhynchus keta*) and sturgeon (*Acipenser transmontanus*) caviar heated by radio frequency. J Food Prot 67:1848–1854

Altieri C, Speranza B, Del Nobile MA et al (2005) Suitability of bifidobacteria and thymol as biopreservatives in extending the shelf life of fresh packed plaice fillets. J Appl Microbiol 99:1294–1302

Alves VF, De Martinis EC, Destro MT et al (2005) Antilisterial activity of a *Carnobacterium piscicola* isolated from Brazilian smoked fish (surubim [*Pseudoplatystoma* sp.]) and its activity against a persistent strain of *Listeria monocytogenes* isolated from surubim. J Food Prot 68:2068–2077

Ananou S, Zentar H, Martínez-Bueno M et al (2014) The impact of enterocin AS-48 on the shelf-life and safety of sardines (*Sardina pilchardus*) under different storage conditions. Food Microbiol 44:185–195

Arlindo S, Calo P, Franco C et al (2006) Single nucleotide polymorphism analysis of the enterocin P structural gene of *Enterococcus faecium* strains isolated from nonfermented animal foods. Mol Nutr Food Res 50:1229–1238

Basti AA, Misaghi A, Salehi TZ et al (2006) Bacterial pathogens in fresh, smoked and salted Iranian fish. Food Control 17:183–188

Bigwood T, Hudson JA, Cooney J et al (2012) Inhibition of *Listeria monocytogenes* by *Enterococcus mundtii* isolated from soil. Food Microbiol 32:354–360

Budu-Amoako E, Ablett RF, Harris J et al (1999) Combined effect of nisin and moderate heat on destruction of *Listeria monocytogenes* in cold-pack lobster meat. J Food Prot 62:46–50

Calo-Mata P, Arlindo S, Boehme K et al (2008) Current applications and future trends of lactic acid bacteria and their bacteriocins for the biopreservation of aquatic food products. Food Bioprocess Technol 1:43–63

Campos CA, Rodriguez O, Calo-Mata P et al (2006) Preliminary characterization of bacteriocins from *Lactococcus lactis*, *Enterococcus faecium* and *Enterococcus mundtii* strains isolated from turbot (*Psetta maxima*). Food Res Int 39:356–364

Cardinal M, Gunnlaugsdottir H, Bjoernevik M et al (2004) Sensory characteristics of cold-smoked Atlantic salmon (*Salmo salar*) from European market and relationships with chemical, physical and microbiological measurements. Food Res Int 37:181–193

Chahad OB, El Bour M, Calo-Mata P et al (2012) Discovery of novel biopreservation agents with inhibitory effects on growth of food-borne pathogens and their application to seafood products. Res Microbiol 163:44–54

Chen H, Hoover DG (2003) Bacteriocins and their food applications. Comp Rev Food Sci Food Safety 2:82–100

Datta S, Janes ME, Xue QG et al (2008) Control of *Listeria monocytogenes* and *Salmonella anatum* on the surface of smoked salmon coated with calcium alginate coating containing oyster lysozyme and nisin. J Food Sci 73:M67–M71

Degnan AJ, Kaspar CW, Otwell WS et al (1994) Evaluation of lactic acid bacterium fermentation products and food-grade chemicals to control *Listeria monocytogenes* in blue crab (*Callinectes sapidus*) meat. Appl Environ Microbiol 60:3198–3203

Dicks LMT, Todorov SD, van der Merwe MP (2006) Preservation of fish spread with enterocins 1071A and 1071B, two antimicrobial peptides produced by *Enterococcus faecalis* BFE 1071. J Food Safety 26:173–183

Diop MB, Dubois-Dauphin R, Destain J et al (2009) Use of a nisin-producing starter culture of *Lactococcus lactis* subsp, *lactis* to improve traditional fish fermentation in Senegal. J Food Prot 72:1930–1934

dos Reis FB, de Souza VM, Thomaz MR et al (2011) Use of *Carnobacterium maltaromaticum* cultures and hydroalcoholic extract of *Lippia sidoides* Cham. against *Listeria monocytogenes* in fish model systems. Int J Food Microbiol 146:228–234

Drider D, Fimland G, Héchard Y et al (2006) The continuing story of class IIa bacteriocins. Microbiol Molec Biol Rev 70:564–582

Duffes F, Corre C, Leroi F et al (1999a) Inhibition of *Listeria monocytogenes* by in situ produced and semipurified bacteriocins of *Carnobacterium* spp. on vacuum-packed, refrigerated cold-smoked salmon. J Food Prot 62:1394–1403

Duffes F, Leroi F, Boyaval P et al (1999b) Inhibition of *Listeria monocytogenes* by *Carnobacterium* spp. strains in a simulated cold smoked fish system stored at 4 °C. Int J Food Microbiol 47:33–42

Einarsson H, Lauzon HL (1995) Biopreservation of brined shrimp (*Pandalus borealis*) by bacteriocins from lactic acid bacteria. Appl Environ Microbiol 61:669–676

Galvez A, Lopez RL, Abriouel H et al (2008) Application of bacteriocins in the control of food-borne pathogenic and spoilage bacteria. Crit Rev Biotechnol 28:125–152

Ghalfi H, Allaoui A, Destain J et al (2006) Bacteriocin activity by *Lactobacillus curvatus* CWBI-B28 to inactivate *Listeria monocytogenes* in cold-smoked salmon during 4 °C storage. J Food Prot 69:1066–1071

González-Rodríguez MN, Sanz JJ, Santos JA et al (2002) Numbers and types of microorganisms in vacuum-packed cold smoked freshwater fish at the retail level. Int J Food Microbiol 77:161–168

Gram L, Dalgaard P (2002) Fish spoilage bacteria – problems and solutions. Curr Opin Biotechnol 13:262–266

Hara H, Ohashi Y, Sakurai T et al (2009) Effect of nisin (Nisaplin) on the growth of *Listeria monocytogenes* in Karashi-mentaiko (Red-pepper Seasoned Cod Roe). Shokuhin Eiseigaku Zasshi 50:173–177

Hiwaki H, Ebuchi S, Baba A et al (2007) Incidence of *Listeria monocytogenes* in Karashi-Mentaiko (red-pepper seasoned cod roe) and a model experiment to control the bacterium by food additives. Jpn J Food Microbiol 24:122–129

Hosseini SV, Arlindo S, Böhme K et al (2009) Molecular and probiotic characterization of bacteriocin-producing *Enterococcus faecium* strains isolated from nonfermented animal foods. J Appl Microbiol 107:1392–1403

Katla T, Møretrø T, Aasen IM et al (2001) Inhibition of *Listeria monocytogenes* in cold smoked salmon by addition of sakacin P and/or live *Lactobacillus sakei* cultures. Food Microbiol 18:431–439

Labbé R, Rahmati T (2012) Growth of enterotoxigenic *Bacillus cereus* on salmon (*Oncorhynchus nerka*). J Food Prot 75:1153–1156

Leisner JJ, Laursen BG, Prévost H et al (2007) *Carnobacterium*: positive and negative effects in the environment and in foods. FEMS Microbiol Rev 31:592–613

Mah JH, Ahn JB, Park JH et al (2003) Characterization of biogenic amine-producing microorganisms isolated from Myeolchi-Jeot, Korean salted and fermented anchovy. J Microbiol Biotechnol 13:692–699

Mah JH, Hwang HJ (2009) Inhibition of biogenic amine formation in a salted and fermented anchovy by *Staphylococcus xylosus* as a protective culture. Food Control 20:796–801

Matamoros S, Pilet MF, Gigout F et al (2009) Selection and evaluation of seafood-borne psychrotrophic lactic acid bacteria as inhibitors of pathogenic and spoilage bacteria. Food Microbiol 26:638–644

Montiel R, Bravo D, Medina M (2013) Commercial biopreservatives combined with salt and sugar to control *Listeria monocytogenes* during smoked salmon processing. J Food Prot 76:1463–1465

Neetoo H, Ye M, Chen H et al (2008a) Use of nisin-coated plastic films to control *Listeria monocytogenes* on vacuum-packaged cold-smoked salmon. Int J Food Microbiol 122:8–15

Neetoo H, Ye M, Chen H (2008b) Potential antimicrobials to control *Listeria monocytogenes* in vacuum-packaged cold-smoked salmon pâté and fillets. Int J Food Microbiol 123:220–227

Nilsson L, Hansen TB, Garrido P et al (2005) Growth inhibition of *Listeria monocytogenes* by a nonbacteriocinogenic *Carnobacterium piscicola*. J Appl Microbiol 98:172–183

Nilsson L, Christiansen NYY, Jorgensen JN et al (2004) The contribution of bacteriocin to inhibition of *Listeria monocytogenes* by *Carnobacterium piscicola* strains in cold-smoked salmon systems. J Appl Microbiol 96:133–143

Nilsson L, Huss HH, Gram L (1997) Inhibition of *Listeria monocytogenes* on cold-smoked salmon by nisin and carbon dioxide atmosphere. Int J Food Microbiol 38:217–227

Nykanen A, Weckman K, Lapvetelainen A (2000) Synergistic inhibition of *Listeria monocytogenes* on cold-smoked rainbow trout by nisin and sodium lactate. Int J Food Microbiol 61:63–72

O'Sullivan L, Ross RP, Hill C (2002) Potential of bacteriocin-producing lactic acid bacteria for improvements in food safety and quality. Biochimie 84:593–604

Paludan-Müller C, Dalgaard P, Huss HH et al (1998) Evaluation of the role of *Carnobacterium piscicola* in spoilage of vacuum and modified atmosphere-packed-smoked salmon stored at 5°C. Int J Food Microbiol 39:155–166

Pao S, Ettinger MR, Khalid MF, Reid AO, Nerrie BL (2008) Microbial quality of raw aquacultured fish fillets procured from internet and local retail markets. J Food Prot 71:1544–1549

Pinto AL, Fernandes M, Pinto C et al (2009) Characterization of anti-*Listeria* bacteriocins isolated from shellfish: potential antimicrobials to control non-fermented seafood. Int J Food Microbiol 129:50–58

Rihakova J, Belguesmia Y, Petit VW et al (2009) Divercin V41 from gene characterization to food applications: 1998–2008, a decade of solved and unsolved questions. Lett Appl Microbiol 48:1–7

Sarika AR, Lipton AP, Aishwarya MS et al (2012) Isolation of a bacteriocin-producing *Lactococcus lactis* and application of its bacteriocin to manage spoilage bacteria in high-value marine fish under different storage temperatures. Appl Biochem Biotechnol 167:1280–1289

Szabo EA, Cahill ME (1999) Nisin and ALTA 2341 inhibit the growth of *Listeria monocytogenes* on smoked salmon packaged under vacuum or 100 % CO_2. Lett Appl Microbiol 28:373–377

Tahiri I, Desbiens M, Kheadr E et al (2009a) Growth of *Carnobacterium divergens* M35 and production of divergicin M35 in snow crab by-product, a natural-grade medium. LWT-Food Sci Technol 42:624–632

Tahiri I, Desbiens M, Kheadr E et al (2009b) Comparison of different application strategies of divergicin M35 for inactivation of *Listeria monocytogenes* in cold-smoked wild salmon. Food Microbiol 26:783–793

Takahashi H, Kuramoto S, Miya S et al (2011) Use of commercially available antimicrobial compounds for prevention of *Listeria monocytogenes* growth in ready-to-eat minced tuna and salmon roe during shelf life. J Food Prot 74:994–998

Tomé E, Gibbs PA, Teixeira PC (2008) Growth control of *Listeria innocua* 2030c on vacuum-packaged cold-smoked salmon by lactic acid bacteria. Int J Food Microbiol 121:285–294

Tsironi TN, Taoukis PS (2010) Modeling microbial spoilage and quality of gilthead seabream fillets: combined effect of osmotic pretreatment, modified atmosphere packaging, and nisin on shelf life. J Food Sci 75:M243–M251

Uyttendaele M, Busschaert P, Valero et al (2009) Prevalence and challenge tests of *Listeria monocytogenes* in Belgian produced and retailed mayonnaise-based deli-salads, cooked meat products and smoked fish between 2005 and 2007. Int J Food Microbiol 133:94–104

Valenzuela AS, Benomar N, Abriouel H et al (2010) Isolation and identification of *Enterococcus faecium* from seafoods: antimicrobial resistance and production of bacteriocin-like substances. Food Microbiol 27:955–961

Vaz-Velho M, Todorov S, Ribeiro J et al (2005) Growth control of *Listeria innocua* 2030c during processing and storage of cold-smoked salmon-trout by *Carnobacterium divergens* V41 culture and supernatant. Food Control 16:541–549

Vescovo M, Scolari G, Zacconi C (2006) Inhibition of *Listeria innocua* growth by antimicrobial-producing lactic acid cultures in vacuum-packed cold-smoked salmon. Food Microbiol 23:689–693

Wan Norhana MN, Poole SE, Deeth HC (2012) Effects of nisin, EDTA and salts of organic acids on *Listeria monocytogenes*, *Salmonella* and native microflora on fresh vacuum packaged shrimps stored at 4 °C. Food Microbiol 31:43–50

Yamazaki K, Suzuki M, Kawai Y et al (2003) Inhibition of *Listeria monocytogenes* in cold-smoked salmon by *Carnobacterium piscicola* CS526 isolated from frozen surimi. J Food Prot 66:1420–1425

Ye M, Neetoo H, Chen H (2008) Effectiveness of chitosan-coated plastic films incorporating antimicrobials in inhibition of *Listeria monocytogenes* on cold-smoked salmon. Int J Food Microbiol 127:235–240

Yücel N, Balci S (2010) Prevalence of *Listeria*, *Aeromonas*, and *Vibrio* species in fish used for human consumption in Turkey. J Food Prot 73:380–384

Zuckerman H, Ben Avraham R (2002) Control of growth of *L. monocytogenes* in fresh salmon using Microgard and nisin. Lebensmittel-Wissenschaft und-Technol 35:543–548

Chapter 8
Biopreservation of Vegetable Foods

8.1 Application of Bacteriocins

8.1.1 Fresh Produce

Fresh produce products can become contaminated with human pathogenic bacteria from different sources (such as manure, irrigation water, insects, and during harvesting and other process operations), and have been implicated in a number of outbreaks (Lynch et al. 2009). Several bacteriocin preparations (such as nisin, pediocin, or enterocin AS-48) have been assayed for inactivation of foodborne pathogenic or toxinogenic bacteria (such as *Listeria monocytogenes*, *Bacillus cereus*, and *Bacillus weihenstephanensis*, *Escherichia coli*, *Salmonella* and other enterobacteria) on the surfaces of fresh-cut vegetables and on sprouted seeds (Galvez et al. 2008; Randazzo et al. 2009; Abriouel et al. 2010) (Table 8.1). Bacteriocin treatments have also been proposed for decontamination of whole fruit surfaces, and to avoid transmission of pathogenic bacteria from fruit surfaces to processed fruits (Ukuku et al. 2005; Silveira et al. 2008), and to decrease bacterial survival of bacteria on sliced fruit surfaces during storage.

Several studies have reported inactivation of human pathogenic bacteria in salads. Torriani et al. (1997) reported that salad vegetables treated with culture supernatant of a *Lactobacillus casei* strain reduced the coliform count, but the presence of a bacteriocin was not confirmed. Randazzo et al. (2009) tested the effect of bacteriocin RUC9 from a wild strain of *Lactococcus lactis* (previously isolated from minimally processed mixed salads) in minimally processed iceberg lettuce samples artificially inoculated with a wild strain of *L. monocytogenes* during storage at 4 °C, in comparison with commercial nisin. None of the bacteriocin treatments completely eliminate the pathogen on the produce, but RUC9 treatment achieved a greater reduction of *L. monocytogenes* viable counts after 7 days of storage at 4 °C compared to nisin (2.7 log units versus 1 log unit). It was suggested that treatment with RUC9 bacteriocin could be used as sanitizer to improve microbial safety and

© The Author(s) 2014
A. Gálvez et al., *Food Biopreservation*, SpringerBriefs in Food, Health, and Nutrition, DOI 10.1007/978-1-4939-2029-7_8

Table 8.1 Example applications of bacteriocin and bacteriocin-producing LAB in vegetable foods and beverages

Bacteriocin treatment	Effect(s)	Reference(s)
Nisin Z, coagulin, nisin:coagulin cocktail	Reduced viable cell counts of *L. monocytogenes* on fresh-cut iceberg lettuce stored in microperforated plastic bags	Allende et al. 2007
Enterocin AS48 washing treatments alone or in combination with other antimicrobials	Inactivation of *L. monocytogenes, B. cereus, B. weihenstephanensis* and *Enterobacteria* on sprouts	Abriouel et al. 2010
Nisin in combination with of hydrogen peroxide, sodium lactate and citric acid as a sanitizer	Decontamination of whole cantaloupe and honeydew melon surfaces. Prevented transfer of *L. monocytogenes* and *E. coli* to fresh cut pieces	Ukuku et al. 2005
Nisin combination with lysozyme and pulsed electric fields (PEF)	Inactivation of *Salmonella* Typhimurium in orange juice	Liang et al. 2002
Enterocin AS-48 alone or in combination with PEF, chelators, or heat	Inactivation of pathogenic and spoilage bacteria in fruit juices	Abriouel et al. 2010
Nisin and high pressure homogeneization	Improved inactivation of *L. monocytogenes* in juice	Pathanibul et al. 2009
Nisin and high hydrostatic pressure	Reduction of aerobic mesophlic microbiota in cucumber juice	Zhao et al. 2014
Nisin	Prevented spoilage caused by non-aciduric and aciduric spore formers in canned foods and in other foods	Thomas et al. 2000, 2001
Enterocin AS-48	Inactivation of endospore formers in boiled rice, purees, and canned vegetables	Abriouel et al. 2010
Nisin	Inactivation of wine LAB at lower sulphite concentrations	Rojo-Bezares et al. 2007
Pediocin PD-1	Control of *O. oeni* in wines	Bauer et al. 2003
Enterocins L50A and L50B	Inhibition of beer spoilage LAB in worts and lager beers	Basanta et al. 2008
Plantaricin-producing starter culture	Improved microbiological control of table olives fermentation	Vega Leal-Sánchez et al. 2003
Kimchicin-producing *Leuconostoc citreum*	Inhibition of foodborne pathogens in kimchi	Chang and Chang 2011
LactiGuard™ lactic acid bacteria	Inhibition of *E. coli* O157:H7 and *C. sporogenes* in spinach	Brown et al. 2011
Bacteriocin-producing strains	Inhibition of rope-forming bacilli in breads	Settanni and Corsetti 2008
	Enhanced competitiveness of strains in fermented doughs	
Bacteriocin-producing *L. plantarum*	Decreased survival of *B. cereus, E. coli* O157:H7 and *S. enterica* in millet gruels	Sánchez Valenzuela et al. 2008

to reduce the chemical treatment in vegetable processing. Allende et al. (2007) tested bacteriocin preparations (nisin Z, coagulin and a nisin:coagulin cocktail) produced by cultivation of selected LAB strains on a lettuce extract in fresh-cut Iceberg lettuce stored in microperforated plastic bags. The applied bacteriocin extracts reduced viable cell counts of *L. monocytogenes*, but did not prevent further growth of survivors during refrigeration storage of samples.

Nisin and pediocin individually or in combination with sodium lactate, potassium sorbate, phytic acid, and citric acid were tested as possible sanitizer treatments for reducing the population of *L. monocytogenes* on cabbage, broccoli, and mung bean sprouts (Bari et al. 2005). After 1-min wash, the combination treatments nisin-phytic acid and nisin-pediocin-phytic acid caused significant reductions of *L. monocytogenes* on cabbage and broccoli but not on mung bean sprouts. Pediocin treatment alone or in combination with any of the organic acid tested was more effective in reducing *L. monocytogenes* populations than the nisin treatment alone (Bari et al. 2005).

Bennik et al. (1999) applied a solution containing 200 AU/ml of mundticin ATO6 (from *Enterococcus mundtii* ATO6) on mung bean sprouts by dipping or coating with an alginate film. Mundticin treatments reduced the levels of *L. monocytogenes* by ca. 2 log cycles after the treatments. However, the bacteriocin treatment had no effect on growth of the listeriae during storage of the treated sprouts under a modified atmosphere (MA) at 8 °C. In another study, alfalfa and soybean sprouts artificially contaminated with *L. monocytogenes* were treated by immersion for 5 min in solutions containing enterocin AS-48 of 5.0, 12.5 and 25 µg/ml. Best results were obtained for samples treated with 25 µg/ml and stored right after treatments at temperatures of 6 or 15 °C, achieving reductions in viable counts of 2.0–2.4 log cycles (Cobo Molinos et al. 2005). In alfalfa and soybean sprouts, the 25 µg/ml bacteriocin treatment reduced viable counts below detection levels and prevented overgrowth of the listeria for a 7-days storage period at temperatures of 6, 15, and 22 °C. However, the same bacteriocin treatment provided more heterogeneous results in green asparagus and failed to inhibit overgrowth of the listeria during storage. The listeridical effect on green asparagus was strongly potentiated when the bacteriocin was applied in combined treatments (like, for example, with lactic acid, trisodium trimetaphosphate, n-propyl-p-hydroxybenzoate, peracetic acid, or sodium hypochlorite). In sprouts and green asparagus artificially contaminated with *B. cereus* and *B. weihenstephanensis*, washing treatments with enterocin AS-48 (25 µg/ml) reduced viable cell counts by up to 2.4 log cycles in samples stored at 6 °C, but not at 15 or 22 °C (Cobo Molinos et al. 2008a). Microbial inactivation improved when the bacteriocin was tested in combination with other antimicrobials (sodium hypochlorite, peracetic acid, polyphosphoric acid or hydrocinnamic acid). For inactivation of *S. enterica* on sprouts, the most effective treatments consisted of solutions containing enterocin AS-48 (25 µg/ml) with lactic acid or polyphosphoric acid (Cobo Molinos et al. 2008b). The combinations of enterocin AS-48 (25 µg/ml) and polyphosphoric acid in a concentration range of 0.1–2.0 % significantly reduced or inhibited growth of the populations of *S. enterica* as well as other Gram-negative bacteria (*E. coli* O157:H7, *Shigella* spp., *Enterobacter aerogenes*, *Yersinia enterocolitica*, *Aeromonas hydrophila* and *Pseudomonas fluorescens*) in sprout samples stored at 6 and 15 °C.

Fresh-cut fruits have been involved in a number of outbreaks due to cross-contamination with human pathogenic bacteria. Nisin reduced *L. monocytogenes* populations on honeydew melon slices and apple slices (Leverentz et al. 2003). Combination treatments containing nisin-sodium lactate, nisin-potassium sorbate and nisin-sodium lactate-potassium sorbate achieved significant reductions of *Salmonella* directly inoculated onto fresh-cut cantaloupe pieces (Ukuku and Fett 2004). Decontamination of whole cantaloupe and honeydew melon surfaces with a combination of hydrogen peroxide, nisin, sodium lactate and citric acid as a sanitizer prevented further transfer of the inoculated *L. monocytogenes* and *E. coli* to fresh cut pieces (Ukuku et al. 2005). In another study, whole fruit pieces and sliced fruits were decontaminated with an enterocin AS-48 solution (25 μg/ml) (Cobo Molinos et al. 2008c). In the absence of treatments, it was found that *L. monocytogenes* was able to multiply in the less acidic sliced (such as pear, kiwi, melon or watermelon). Bacteriocin treatments significantly inhibited or completely inactivated *L. monocytogenes* in strawberries, raspberries, and blackberries stored at 15 and 22 °C for up to 2 days and in blackberries and strawberries at 6 °C for up to 7 days. Washing treatments also reduced viable counts in sliced melon, watermelon, pear, and kiwi but did not avoid proliferation of survivors during storage at 15 and 22 °C. However, it was reported that combinations of enterocin AS-48 with carvacrol or with *n*-propyl *p*-hydroxybenzoate avoided overgrowth of listeria on sliced melon during storage at 22 °C. Some of the combined treatments proposed could find industrial applications, especially in added-value food products such as those intended for consumption by the elderly, immunocompromised people, or debilitated hospital patients.

8.1.2   Fruit Juices and Drinks

Bacteriocins could be exploited for biopreservation of in fruit and vegetable juices and drinks. These substrates exhibit certain features that may enhance bacteriocin activity and stability, e.g.: (1) improved diffusion of bacteriocin molecules (compared to solid substrates), (2) a low fat content in general, minimizing bacteriocin adsorption to hydrophobic food components, (3) an acidic pH, which in general facilitates bacteriocin solubility and activity and (4), they usually contain organic acids and other bioactive molecules which may potentiate bacteriocin activity. In fruit juices and drinks, bacteriocin addition (nisin, enterocins) has been proposed for inactivation of endospore-forming bacteria causing spoilage such as *Alicyclobacillus acidoterrestris* and thermophilic spore formers such as *Geobacillus stearothermophilus*. Bacteriocin addition may also be useful for microbial inactivation of bacteria causing ropiness (such as exopolysaccharide-producing *Bacillus licheniformis*, pediococci and lactobacilli), as well as acrolein-producing bacteria. While fruit juices and drinks usually have a pH that is too low for proliferation of foodborne pathogenic bacteria, some less acidic juices and drinks can support bacterial growth. Inactivation of foodborne pathogens (*L. monocytogenes*, *B. cereus*,

Staphylococcus aureus) by enterocins has been reported in lettuce juices, soy milk, and sport and energy drinks with lower acidity (Galvez et al. 2008; Abriouel et al. 2010). Since freshly-made fruit juices have been implicated in transmission of enteric pathogens, bacteriocins (such as nisin and enterocin AS-48) have been tested in combination with other agents to increase the bacterial outer membrane permeability. The combined treatments of bacteriocins and PEF greatly increased the bactericidal effects and decreased the risks of survivor proliferation in the treated samples (Liang et al. 2002; Mosqueda-Melgar et al. 2008).

In fruit juices, the thermophilic endospore former *A. acidoterrestris* can withstand pasteurisation temperatures commonly applied during food processing, and spoil freshly-made juices as well as processed juices. Even moderate growth of this bacterium can confer an unpleasant medicinal taste to fruit juices, due to the production of guaiacol. Some bacteriocins (mainly nisin and enterocin AS-48) have been suggested as possible hurdles against this bacterium. Nisin (1.25–100 IU/ml) was able to inhibit *A. acidoterrestris* in orange juice, grapefruit juice and apple juice (Komitopoulou et al. 1999) as well as in orange and fruit-mixed drinks (Yamazaki et al. 2000). Nisin was also able to inhibit spore germination at 25–50 IU/ml in orange and mixed fruit drinks, but not by 600 IU/ml in clear apple juice, probably because of a competitive effect of phenols (Yamazaki et al. 2000). Addition of nisin to orange juice (0, 50, 75, and 100 IU of nisin/ml juice) increased the thermal death of *A. acidoterrestris* spores, with reported decrease in the D value up to 27 % heat resistance as the nisin concentration was increased (Peña et al. 2009). Enterocin AS-48, added at low concentrations of 2.5 µg/ml in fruit juices artificially contaminated with vegetative cells and endospores of *A. acidoterrestris* caused complete bacterial inactivation and afforded protection for up to 14 days in freshly made orange and apple juices and for up to 60–90 days in several commercial fruit juices (Grande et al. 2005). Electron microscopy examination of bacteriocin-treated vegetative cells revealed substantial cell damage and bacterial lysis. Treatment of endospores with enterocin AS-48 caused inhibition of germination and disorganisation of endospore structure.

Another heat-resistant bacterium in fruit juices is the non-sporeformer *Propionibacterium cyclohexanicum*, implicated in the spoilage of orange juice (Kusano et al. 1997). One study showed that addition of nisin (500 and 1,000 IU/ml) to orange juice significantly reduced the viable population of *P. cyclohexanicum* for up to 15 days, but did not prevent regrowth of the bacterium during higher storage periods (Walker and Phillips 2008).

Fruit juices may also be spoiled by bacteria producing exopolysaccharides (EPS), acrolein, or simply by growth causing turbidity. In apple juice and apple ciders, added enterocin AS-48 (2.5–5 µg/ml) was very effective against EPS-producing bacterial strains (including *B. licheniformis* LMG 19409, *Lactobacillus collinoides*, *Lactobacillus diolivorans* and *Pediococcus parvulus*) as well as 3-hydroxypropionaldehyde -producing *L. collinoides* strains (Grande et al. 2006b; Martínez-Viedma et al. 2008a). In coconut juice and coconut milk, addition of enterocin AS-48 (at a final concentration as low as 1.75 µg/ml) completely suppressed *G. stearothermophilus* for at least 30 days of incubation at 45 °C (Martínez-Viedma et al. 2009b).

Many different reports have indicated that unpasteurized fruit juices have been involved in outbreaks due to contamination with pathogenic enteric bacteria such as *E. coli* and *Salmonella enterica*. Since bacteriocins in general are not active on Gram-negative bacteria due to the protective barrier of the bacterial outer membrane, different trials have been carried out in which bacteriocins were tested in combination with outer membrane-permeabilising treatments or antimicrobial agents in order to increase microbial inactivation. In apple juice, a combination of nisin and cinnamon improved the inactivation of *Salmonella* Typhimurium and *E. coli* O157:H7, therefore enhancing the safety of the juice (Yuste and Fung 2004). Nisin (300 IU) in combination with EDTA (20 mM) caused a decline in the populations of *E. coli* O157:H7, *Salmonella*, and *L. monocytogenes* in apple cider, suggesting possible addition of this preparation to freshly prepared apple cider to enhance its microbial safety and prevent costly recalls (Ukuku et al. 2009). In apple juice, it was shown that *E. coli* O157:H7 cells sublethally injured by outer membrane permeabilizing treatments (EDTA, sodium tripolyphosphate, pH 5.0, and moderate heat) became sensitive to enterocin AS-48. Highest bactericidal activity (up to to 8.1 log cycles inactivation) was observed when the bacteriocin was applied in multiple treatments (Ananou et al. 2005).

Non-thermal food processing technologies are gaining interest in food preservation. Treatments with high-intensity pulsed electric fields (PEF) are quite effective in microbial inactivation in pumpable substrates such as fruit juices while having little or no effect on organoleptic and nutritional properties (Martín-Belloso and Elez-Martínez 2005; Mittal and Griffiths 2005). Therefore, several studies have explored the synergistic effects of PEF treatments in fruit juices. Addition of nisin (2 %, wt/vol) to freshly squeezed, unpasteurized, and preservative-free apple juice followed by application of a PEF treatment (80 kV/cm, 10 pulses, 42 °C) caused a greater reduction in *E. coli* O157:H7 cell counts of more than 3 log cycles compared to application of PEF treatment alone (Iu et al. 2001).

Nisin and/or lysozyme in combination with PEF treatments were studied on *S.* Typhimurium in pasteurized and freshly squeezed orange juice (Liang et al. 2002). It was found that increasing the treatment temperature to 45 °C or above was critical for inactivation of *Salmonella* by PEF, suggesting a synergistic effect of heat in the inactivation process. Application of PEF treatments (90 kV/cm, 30 pulses, 45 °C) in the presence of nisin (100 U/ml of orange juice), lysozyme (2,400 U/ml), or a mixture of nisin (27.5 U/ml) and lysozyme (690 U/ml), achieved much greater cell inactivation compared with the single treatments. The most effective PEF treatment in juice was obtained using a combination of nisin and lysozyme, achieving a ca. 6.5 log cycle-reduction in viable counts of *Salmonella* (Liang et al. 2002).

Enterocin AS-48 (60 μg/ml) in combination with PEF treatments (35 kV/cm, 1,000 μs) at 40 °C was tested against *S. enterica* cells in apple juice. The combined treatments decreased viable counts of the pathogen by 4.5-log cycles, while treatment with bacteriocin alone had no effect (Martínez-Viedma et al. 2008b). PEF treatment of apple juice combined with enterocin AS-48 (at a subinhibitory concentration of 2 mg/l) also showed higher efficacy against the exopolysaccharide-producing, fruit juice-spoilage strain *L. diolivorans* 29 compared to PEF and enterocin alone (Martinez-Viedma et al. 2009c). Bacteriocin addition to apple juice also served as an additional hurdle after treatments, preventing regrowth of survivors during storage of treated samples for at least 15 days at 4 and 22 °C (Martinez-Viedma et al. 2009c).

The combinations of bacteriocins with PEF treatments have been explored with the aims to achieve higher reductions of total microbial loads and to prevent or retard food spoilage. In one study carried out on freshly squeezed apple cider, it was shown that inactivation of naturally occurring microorganisms (yeast and molds) increased by 1.1–1.8 logs upon application of PEF treatment in combination with a nisin/lysozyme mixture (Liang et al. 2006). A similar combined treatment was tested for inactivation of naturally occurring spoilage microorganisms in red and white grape juices, achieving reductions in viable bacterial counts up to 5.9 logs (Wu et al. 2005). Also, addition of nisin (100 U/ml) in tomato juice before application of a PEF treatment (80 kV/cm, 20 pulses, 50 °C) was reported to stably reduce cell counts by about 4.4 log units during 28 days of storage at 4 °C, with the advantage that vitamin C content was not affected –as otherwise would happen after application of more intense heat treatments (Nguyen and Mittal 2007).

High pressure homogeneization (HPH) sensitises *E. coli* cells to antibacterial peptides and enzymes (Diels et al. 2004, 2005). In one study, Pathanibul et al. (2009) tested the effect of HPH (0–350 MPa) on *E. coli* and *Listeria innocua* cells in apple or carrot juice. Addition of nisin (10 IU) in combination with HPH increased the bactericidal effect against *L. innocua* in apple juice as well as in carrot juice, reducing to some extent the intensity of the high pressure treatment for inactivation of this bacterium (Pathanibul et al. 2009). However, inactivation of *E. coli* cells in juices did not improve by addition of nisin in combination with HPH. Bacteriocins could also improve the efficacy of high hydrostatic pressure (HHP) treatments in vegetable foods. In one study, it was found that the combination of nisin with HHP had a synergistic effect on the inactivation of total plate count in cucumber juice (Zhao et al. 2014).

8.1.3 Ready-to-Eat, Processed, and Canned Vegetable Foods

Deli-type salads are very popular ready-to-eat vegetable foods. These food products often contain mixtures of cooked and/or uncooked vegetables (e.g. potatoes, tomatoes, olives, peas, carrots, lettuce or cabbage) and other ingredients (e.g. ham, chicken, tuna, egg, or seafood) blended with mayonnaise or salad dressing. Extensive handling during preparation by foodservice personnel, cross contamination, potential abusive temperatures during storage, and the lack of heat-treatment before consumption are factors that may compromise the microbiological safety of these foods. Because of the low numbers of competing microbiota as a result of cooking steps, foodborne pathogens can easily proliferate in salads (Jay et al. 2005). Transmission of bacterial pathogens such as *S. enterica* and *L. monocytogenes* by deli salads is a public health concern (Unicomb et al. 2003; Mokhtari et al. 2006).

Addition of nisin in deli-type salads reduced the concentrations of viable *Listeria* following bacteriocin treatment, but it did not prevent overgrowth of survivors during storage (Schillinger et al. 2001). In vegetable salads (containing a mayonnaise-blended mixture of ingredients such as boiled potato, carrots, peas, olives, egg and tuna) complete inactivation of *L. monocytogenes* population was achieved by adding enterocin AS-48 at 60 µg/g singly or at 30 µg/g in combinations with a variety of antimicrobial substances such as essential oils or their bioactive compounds, and chemical preservatives (Cobo Molinos et al. 2009a). Synergistic effects were also reported in salads against a cocktail of *S. enterica* serovars for AS-48 and *p*-hydroxybenzoic acid methyl esther or 2-nitropropanol (Cobo Molinos et al. 2009b). In vegetable sauces, inactivation of *S. aureus* improved considerably when combinations of enterocin AS-48 and hydrocinnamic acid or carvacrol were added (Grande et al. 2007a).

In cooked vegetable foods, most vegetative cells can be inactivated during the heat processing. However, endospores surviving heat treatments can proliferate in the finished product unless additional hurdles (such as bacteriocins) are included. Enterocin AS-48 added in a concentration range of 20–35 µg/g to boiled rice and in a commercial infant rice-based gruel dissolved in whole milk and inoculated with vegetative cells and endospores of *B. cereus*, completely suppressed the bacilli during storage in a temperature range of 6–37 °C for up to 15 days and prevented enterotoxin production (Grande et al. 2006b). Bacteriocin activity was improved by adding sodium lactate, decreasing the bacteriocin concentration to 8–16 µg/g without compromising inactivation of the bacilli. Application of AS-48 in combination with heat treatments decreased the thermal death D values for endospores (Grande et al. 2006b). In desserts and bakery ingredients, the efficacy of AS-48 (added in a concentration range of 5–50 µg/g) against *S. aureus*, *B. cereus*, and *L. monocytogenes* depended to a great extent on the food substrate and the target bacteria (Martínez-Viedma et al. 2009a, b, c). Bacteriocin activity in chocolate cream increased markedly when tested in combination with eugenol, 2-nitropropanol or Nisaplin (Martínez-Viedma et al. 2009c).

L. monocytogenes has been reported to grow on tofu stored at refrigeration temperatures (Schillinger et al. 2001). Nisin showed limited efficacy against *L. monocytogenes* Scott A in tofu. Following an initial reduction of viable counts by nisin, regrowth of survivors was observed during further incubation (Schillinger et al. 2001). To improve microbial inactivation, nisin was tested in combination with bacteriocinogenic strains *Enterococcus faecium* BFE 900-6a or *L. lactis* BFE 902 as protective cultures, resulting in a complete suppression of listerial growth in home-made tofu stored at 10 °C for 1 week (Schillinger et al. 2001).

Canning and cooking processes of vegetables destroy most of the vegetative bacterial forms. Yet, due to the high thermal resistance of endospores and the frequent endospore contamination of raw materials, endospore-forming bacteria represent the main risk for spoilage of foods prepared in this way. Additional hurdles such as refrigeration, acidification, addition of salt or chemical preservatives are often required to avoid proliferation of sporeformers in the processed products. Several studies support the practical application of bacteriocins in this category of foodstuffs to inhibit endospore outgrowth and also to increase the efficacy of thermal treatments against endospores. Incorporation of nisin in canned vegetables can prevent spoilage caused by non-aciduric (*Bacillus stearothermophilus* and *Clostridium thermosaccharolyticum*) as well as by aciduric (*Clostridium pasteurianum, Bacillus macerans, Bacillus coagulans*) spore formers (Thomas et al. 2000). Nisin has also been reported to be an effective preservative in fresh pasteurized "home-made"-type soups (Thomas et al. 2000) and in the control of *Bacillus* and *Clostridium* in cooked potato products (Thomas et al. 2002). For example, in pasteurised, vacuum-packaged mashed potatoes inoculated with a cocktail of *Clostridium sporogenes* and *Clostridium tyrobutyricum* spores, addition of nisin prevented bacterial growth and extended the shelf life of the mashed potatoes by at least 30 days (Thomas et al. 2002). Similar results were reported following nisin addition in trials involving a cocktail of *B. cereus* and *Bacillus subtilis* strains (Thomas et al. 2002). Nisin could be applied singly or in combination with other bacteriocins, as shown in sous vide mushrooms, in which addition of a nisin-pediocin mixture prevented outgrowth of *B. subtilis* spores (Cabo et al. 2009).

In cooked vegetables (such as cooked potato products, sous-vide mushrooms, "home-made"-type soups, purees, or cooked rice foods) and in canned vegetables (such as canned tomato, peas, corn, etc.), addition of bacteriocins (such as nisin, nisin-pediocin combination, or enterocin AS-48) has been proposed as a way to inhibit endospore outgrowth and production of enterotoxins (such as *B. cereus* or *Clostridium botulinum* toxins) during storage and/or to increase the efficacy of thermal treatments against endospores (Thomas et al. 2000; Galvez et al. 2008; Cabo et al. 2009; Abriouel et al. 2010). Incorporation of bacteriocins in canned vegetables can be an effective hurdle to prevent spoilage caused by non-aciduric as well as by aciduric spore formers.

In vegetable food products processed by heat such as purees and canned vegetables, enterocin AS-48 was tested against endospore-forming bacteria. In one study, by adding AS-48 (10 mg/l), *B. cereus* LWL1 was completely inhibited in six vegetable products (natural vegetable cream, asparagus cream, traditional soup, homemade-style traditional soup, vegetable soup, and vichyssoise) for up to 30 days in samples stored at 6, 15, and 22 °C. Other *Bacillus* and *Paenibacillus* species and strains isolated from purées showed variable degrees of inactivation by enterocin AS-48, requiring bacteriocin concentrations up to 50 μg /ml (Grande et al. 2007b). Antimicrobial activity on a cocktail of bacilli increased considerably in combination with phenolic compounds (carvacrol, eugenol, geraniol, and hydrocinnamic acid) (Grande et al. 2007b). In canned foods such as tomato paste, syrup from canned peaches, and juice from canned pineapple, *B. coagulans* (responsible for flat sour spoilage) was inhibited by added enterocin AS-48 (6 mg /l) for at least 15 days of storage at 37 °C (Lucas et al. 2006). Added bacteriocin also increased heat inactivation of *B. coagulans* endospores.

In canned corn and peas, addition of enterocin AS-48 (7 μg/g) inactivated *G. stearothermophilus* cells for at least 30 days at a temperature of 45 °C simulating tropical conditions (Martínez Viedma et al. 2010a). Remarkably, AS-48 strongly adsorbed to bacterial endospores, inhibiting endospore outgrowth. Enterocin EJ97

(2–4 AU/ml) could also inactivate *G. stearothermophilus* vegetative cells in canned corn and peas, and it increased the efficacy of heat treatments on endospores (Martínez Viedma et al. 2010a). Enterocin EJ97 immobilized by coating in polythene films in combination with EDTA reduced the concentrations of viable *B. coagulans* cells canned corn and peas stored at 4 °C (Martínez Viedma et al. 2010b).

8.1.4 Fermented Vegetables and Beverages

Most vegetable fermentations are spontaneous (that is, no starter cultures being added) and rely on the selective growth of the microbiota present in the raw materials as well as microorganisms acquired during handluing a processing, coming from water, equipments, and the processing environment. Addition of bacteriocins (such as nisin) has been proposed as a way to direct the microbiota of vegetable fermentations towards selection of desirable bacteriocin-tolerant or bacteriocin-resistant strains with desirable effects (which in part may be due which due to their homofermentative or heterofermentative traits) while at the same time inhibiting strains causing defects such as overripening (as in the case of kimchi fermentation) or spoilage.

Addition of a nisin preparation to cabbage inoculated with nisin-resistant *Ln. mesenteroides* improved control of the fermentation and delayed growth of the homofermentative LAB (Breidt et al. 1995). In kimchi, nisin was added to control lactobacilli responsible for over-ripening of the product. Nisin addition showed higher growth inhibition of *Lactobacillus* spp. than *Leuconostoc* spp. (Choi and Park 2000).

Ropiness of bread is mainly caused by *B. subtilis*, but *B. licheniformis*, *Bacillus megaterium* and *B. cereus* may also be involved. Rope formation may occur in wheat breads that have not been acidified, or in breads with high concentrations of sugar, fat, or fruits (Beuchat 1997). The application of Nisaplin or nisin-producing lactic acid bacteria in bread production was considered to be ineffective for inhibition of *B. subtilis* and *B. licheniformis* strains (Rosenquist and Hansen 1998). The fermented broth from a *Lactobacillus plantarum* strain producing bacteriocin-like inhibitors was reported to inhibit rope formation by *B. subtilis* in yeast-leavened bread (Valerio et al. 2008). Other bakery products may combine a variety of ingredients (as is the case of refrigerated pizza), making the control of microbial spoilage more difficult. In ham pizza, application of Nisaplin under modified atmosphere packaging significantly increased the product shelf life, due to inhibition of spoilage lactic acid bacteria (Cabo et al. 2001).

In fermented beverages, bacteriocin preparations can be applied against spoilage LAB. In the beer production process, several applications have been proposed for nisin (Ogden and Tubb 1985; Ogden and Waites 1986; Radler 1990; Delves-Broughton et al. 1996; Jespersen and Jakobsen 1996; Thomas et al. 2000): (1) cleaning of the equipment and final cleansing rinse; (2) addition to fermenters to control contamination; (3) increasing the shelf life of uncontaminated beers; (4) reduction of pasteurization regimes, and (5) washing pitching yeasts to eliminate contaminating bacteria, and (6) development of wort bioacidifiying-LAB and/or yeast starter cultures genetically modified to produce bacteriocins. Although nisin activity is limited to Gram-positive bacteria, the sensitivity of the Gram-negative bacterium *Pectinatus frisingensis* to low nisin concentrations was reported in one study (Chihib et al. 1999a). However, selection of a strain resistant to high nisin concentration was also described (Chihib et al. 1999b). Nisin was reported to act synergistically in combination with potassium metabisulphite against wine LAB, and it was proposed that the addition of nisin could be applied in order to reduce the concentrations of sulphur dioxide currently used in the wine industry (Bartowsky 2009; Rojo-Bezares et al. 2007).

Bacteriocins from strains of *Lactobacillus sakei* and *Ln. mesenteroides* isolated from malted barley have been proposed as biological control agents in the brewing industry (Vaughan et al. 2001). Bacteriocins produced by bacterial strains from other sources could also be useful, provided they inhibit target spoilage bacteria in the brewing process. For example, application of thermophilin 110 from *Streptococcus thermophilus* in the brewing industry has been suggested based on its high antimicrobial activity on pediococci (Gilbreth and Somkuti 2005). Enterocins have also been tested for biopreservation of beers. Enterocins L50A and L50B produced by strain *E. faecium* L50 were bactericidal against the most relevant beers spoilage LAB (i.e., *Lactobacillus brevis* and *Pediococcus damnosus*) in a dose- and substrate-dependent manner when challenged in wort, alcoholic and non-alcoholic lager beers at 32 °C. Enterocin addition achieved log reductions of ca 5 log cycles, and no bacterial resistances were detected even after incubation for 6–15 days (Basanta et al. 2008).

Nisin addition is permitted in beer in certain countries, but not in wine. However, nisin has been reported to act synergistically with sulphites against wine LAB (Rojo-Bezares et al. 2007). Nisin addition could aid to reduce the sulphite content in musts before fermentation. One limitation of nisin addition in certain wines would be inhibition of bacteria responsible for the malolactic fermentation. To solve this shortcoming, nisin-resistant strains of *Oenococcus oeni* have been developed that can grow and maintain malolactic wine fermentation in the presence of nisin.

Pediocins may also find applications in wine. Pediocin N5p from *Pediococcus pentosaceus* is resistant to the physico–chemical factors involved in vinification i.e. pH, temperature, ethanol and SO (Strasser de Saad et al. 1995). Application of pediocin PD-1 produced by *P. pentosaceus* isolated from beer, has been proposed in removal of *O. oeni* biofilms from stainless steel surfaces and also to control growth of *O. oeni* in wine (Bauer et al. 2003). PD-1 was the most effective bacteriocin in removal of an established biofilm from stainless steel surfaces in Chardonnay must when compared with nisin and plantaricin 423 (Nel et al. 2002). *Pediococcus acidilactici* J347-29 produced pediocin PA-1 in presence of ethanol and grape must, suggesting its potential biopreservative in winemaking (Díez et al. 2012). The authors tested the effect of pediocin PA-1 alone and in combination with sulphur dioxide and ethanol on the growth of a collection of 53 oenological LAB, 18 acetic acid bacteria and 16 yeast strains. Acetic acid bacteria and yeasts were not inhibited by pediocin PA-1. *O. oeni* was the most sensitive bacterium compared with other wine

LAB (IC50 values of 19 and 312 ng/ml, respectively). Pediocin inhibitory effect was slightly enhanced in the presence of low concentrations of ethanol (6 %). *O. oeni* was also the most sensitive wine LAB when challenged with lacticin 3147 (García-Ruiz et al. 2013). Inhibitory activity of the bacteriocin was strain-dependent, and one *L. casei* strain was remarkably resistant. A synergistic effect of lacticin 3147 and metabisulphite was demonstrated.

Addition of bacteriocins could prevent spoilage of other alcoholic beverages, such as fermented apple ciders (Galvez et al. 2008; Abriouel et al. 2010). Bacteria causing spoilage of apple juice and fermented apple ciders such as exopolysaccharide-producing lactobacilli (*L. collinoides, L. diolivorans* and *P. parvulus* strains) and 3-hydroxypropionaldehyde (3-HPA)-producing *L. collinoides* strains were inhibited efficiently in apple juice and in apple ciders by bacteriocin concentrations in the range of 0.5–5 µg/ml (Grande et al. 2006a; Martínez-Viedma et al. 2008a).

Vegetative cells of the rope-forming strain *B. licheniformis* LMG 19409 (a spoilage strain isolated from spoiled Normand ciders) were rapidly inactivated by bacteriocin addition in fresh-made apple juice and in commercial apple ciders. Although bacterial endospores from *B. licheniformis* were resistant to the bacteriocin, the presence of AS-48 increased the heat sensitivity of endospores. The combination of bacteriocin and moderate heat treatments (85–100 °C) increased the heat inactivation of endospores in cider, decreasing *D* and *z* values (Grande et al. 2006a).

8.2 Bacteriocin-Producing Strains

Inoculation with live cultures is another proposed alternative for inhibition of pathogenic bacteria on fresh produce surfaces (Table 8.1). Bacterial strains (including species of genera such as *Bacillus, Pseudomonas, Enterococcus, Lactococcus, Leuconostoc, Weissella,* or *Lactobacillus*) isolated from raw vegetables may produce antagonistic substances to foodborne pathogens (Galvez et al. 2008; Trias et al. 2008a,b). These strains may be better adapted to vegetable substrates and growth under cold or moderate temperatures. The efficacy of such treatments greatly depends on ecological factors such as the capacity to grow and produce antimicrobials in situ by the protective cultures in competition with resident microbiota.

LactiGuard™ (Guardian Food Technologies, LLC) is defined as "bacterial and bacteriological preparations applied to meat, poultry, pork and produce, namely, fresh fruits and vegetables, to decrease pathogens and improve human food safety". When applied by spray on spinach in combination with water and chlorine as additional hurdles followed by MAP storage 9 days at 4–7 °C, the treatment inhibited *E. coli* O157:H7 and *C. sporogenes* compared to controls with reductions of 1.43 and 1.10 log, respectively (Brown et al. 2011). In freshly harvested spinach, application of the commercial LAB food antimicrobial at 8.0 log CFU/g produced significant ($p < 0.05$) reductions in *E. coli* O157:H7 and *Salmonella* populations on spinach of 1.6 and 1.9 log CFU/g, respectively in the course of aerobic incubation for 12 days at 7 °C (Cálix-Lara et al. 2014).

Fermented vegetables are good sources for isolation of LAB producing bacteriocins. Some examples of antagonistic LAB include *L. lactis* 23 from fermented carrots (Uhlman et al. 1992), *L. plantarum* strains C-11 and C19 from cucumber fermentations (Daeschel et al. 1990; Atrih et al. 1993), *Lactobacillus sake* C2 from traditional Chinese fermented cabbage (Gao et al. 2010), *P. pentosaceus* 05-10 isolated from Sichuan Pickle, a traditionally fermented vegetable product from China (Huang et al. 2009), *L. plantarum* LPCO10 from fermented table olives, (Jimenez-Diaz et al. 1993), *L. plantarum* strains ST23LD and ST341LD from spoiled olive brine (Todorov and Dicks 2005), *Lactobacillus pentosus* B96 from fermenting green olives (Delgado et al. 2005), and *E. faecium* BFE 900 from fermented black olives (Franz et al. 1996). These strains offer potential for investigation as starter or protective cultures in vegetable fermentations, but there are still only limited numbers of studies on this issue.

During cabbage fermentation, inoculation with a paired culture consisting of a nisin-producer *L. lactis* and a nisin-resistant *Ln. mesenteroides* was tested with the purpose of improving the fermentation (Harris et al. 1992a,b). In kimchi, inoculation with a pediocin-producing strain of *P. acidilactici* was reported to successfully achieve inhibition of *L. monocytogenes*, thus improving the product safety (Choi and Beuchat 1994). In a more recent study, kimchi was prepared with *Leuconostoc citreum* GJ7 (producer of the bacteriocin kimchicin GJ7), with the objective of preventing growth and/or survival of foodborne pathogens *E. coli* O157:H7, *Salmonella* Typhi, and *S. aureus* (Chang and Chang 2011). Viable cell reductions of 3.85, 4.45, and 5.19 log CFU/ml were observed 48 h after inoculation. The study concluded that addition of a starter culture capable of producing bacteriocins could serve as a strategy to protect the fermented product from delivering pathogens upon consumption and that the kimchi filtrate itself may be used as a food preservative.

Table olive fermentations are very popular in countries from the Meditarranean region. LAB (mainly *L. plantarum* and *L. pentosus*) together with yeasts are the main bacterial group responsible for these fermentations. These lactobacilli may produce several different bacteriocins (known as plantaricins), and plantaricin genes seem to be widely disseminated (Maldonado et al. 2002). In the Spanish-style process for table olive preparation, green olives are first treated with lye, a treatment that destroys most of the epiphytic microbiota. The lactic fermentation that takes place afterwards is often a slow process that relies mostly on the resident microbiota from the fermentation tanks and manufacturing plant environment. Fast acidification is crucial for proper preservation of olives and inhibition of adventitious microbiota. In some cases, sufficient lactic acid is not produced to warrant product preservation, and spoilage may occur unless exogenous acid is supplemented. The strain *L. plantarum* LPCO10 (plantaricin S and T producer) has been patented for application as a starter culture in the fermentation of table olives and other vegetable foods (Jiménez-Díaz et al. 1993; Ruiz-Barba et al. 1994; Vega Leal-Sánchez et al. 2003). In table olives, inoculation with the plantaricin-producing culture improved the microbiological control of the fermentation process, increased the lactic acid yield and provided a consistent high quality product (Ruiz-Barba et al. 1994; Vega Leal-Sánchez et al. 2003). The starter culture could also be applied to ensure homogeneous and faster fermentations in newly-operating plants that still lack the appropriate resident LAB microbiota. *L. plantarum* is also the main bacterial species in other traditional vegetable fermentations such as capers and Almagro eggplants (Seseña et al. 2004; Pérez Pulido et al. 2005). Therefore, plantaricin-producing strains could also find applications in these (and probably others) fermented vegetables.

Cereals and fermented doughs can also be a good source of LAB strains producers of bacteriocins and bacteriocin-like substances, such as bavaricin A, plantaricin ST31, BLIS C57, amylovorin L and others (Messens et al. 2002; Messens and De Vuyst 2002; Narbutaite et al. 2007; Settanni and Corsetti 2008) as well as antifungal compounds (Valerio et al. 2009; Dalié et al. 2010). Although amylovorin is not active against *Bacillus*, amylovorin production may serve to enhance the competitiveness of the producer strain against other lactobacilli in the fermentation (Messens et al. 2002). After a more direct screening for LAB producing bacteriocin-like antirope activities, two strains were selected (*L. plantarum* E5 and *Ln. mesenteroides* A27) that inhibited ropiness in the bread for more than 15 days (Pepe et al. 2003). According to another study, production of antimicrobial activity by sourdough LAB appears to occur at a low frequency, but the producer strains are active in producing antimicrobial activity under sourdough and bread-making conditions (Corsetti et al. 2004). One *L. lactis* strain isolated from raw barley showed a wider inhibitory spectrum than sourdough LAB (Settanni et al. 2005). This strain was found to produce lacticin 3147, and was shown to produce bacteriocin in situ without interfering with growth of bacteriocin-insensitive *Lactobacillus sanfranciscensis* strains. Furthermore, the production of antimicrobial substances, such as reutericyclin and bacteriocins may enhance the competitiveness of strains in fermented doughs (Gänzle and Vogel 2003; Leroy et al. 2007), being considered a desirable trait for starter culture development (Corsetti and Settanni 2007; De Vuyst et al. 2009). It has been suggested that sourdoughs or cultured broths fermented with bacteriocin-producing lactobacilli could be applied to inhibit rope formation by bacilli in yeast-leavened breads (Menteş et al. 2007; Settanni and Corsetti 2008; Valerio et al. 2008).

Ethnic fermented vegetable foods are good candidates for isolation of antagonistic strains producers of (maybe) new bacteriocins, specifically adapted to the particular fermentation conditions of these foods (Kostinek et al. 2007; Yoon et al. 2008; Ge et al. 2009; Hata et al. 2009; Huang et al. 2009; Tamang et al. 2009; Gao et al. 2010). In one study, the nisin-producer *L. lactis* subsp. *lactis* IFO12007 isolated from miso was used as starter for fermentation of cooked rice and rice koji supplemented with soybean extract (Kato et al. 2001). The producer strain proliferated in the cooked rice and produced enough nisin activity to inhibit *B. subtilis* without causing any adverse effect on growth of *Aspergillus oryzae* during the koji fermentation. Furthermore, a lower salt content could be added to rice miso without compromising the lactic acid fermentation of both rice and soybeans (Kato et al. 2001).

Fermented millet flours are widely consumed in many African countries. Yet, the number of studies carried out on their LAB microbiota and their bacteriocins are limited (Ben Omar et al. 2006, 2008). However, results indicate that LAB strains from these fermented foods may have strong inhibitory activities against foodborne pathogens. In one study, the plantaricin-producing strain *L. plantarum* 2.9 (isolated from ben saalga, a traditional pearl millet fermented food from Burkina Faso) produced strong inhibitory activity in malted millet flour, decreasing the survival of *B. cereus*, *E. coli* O157:H7 and *S. enterica* (Sánchez Valenzuela et al. 2008). The authors of this study suggested that this strain could be applied as a starter culture to improve the microbiological safety of cereal gruels.

Bacteriocin-producing strains can be isolated from raw as well as malted barley. Fermented worts containing bacteriocins could be used to prevent beer spoilage LAB (Vaughan et al. 2005). Bacteriocin production may be a desirable trait for wort bioacidifying LAB starter cultures, enhancing the implantation and proliferation of such strains over spoiling LAB. Development of yeast starter cultures genetically modified to produce bacteriocins has also been suggested, and heterologous production of bacteriocins such as pediocins, leucocins, plantaricins and enterocins by yeasts has been reported (Schoeman et al. 1999; Du Toit and Pretorius 2000; Van Reenen et al. 2002; Gutiérrez et al. 2005; Sánchez et al. 2008; Basanta et al. 2009). The bactericidal yeast strains could be used as starters or protective cultures in the fermentations of brewing, wine and baking processes as biological control agents to inhibit growth of spoilage bacteria.

Several bacteriocinogenic LAB strains have been isolated from wine and wineyards, including species of *L. plantarum*, *O. oeni* and *P. pentosaceus* (Lonvaud-Funel and Joyeux 1993; Strasser de Saad and Manca de Nadra 1993; Navarro et al. 2000; Rojo-Bezares et al. 2007; Knoll et al. 2008; Yanagida et al. 2008). Selected bacteriocin-producing strains could be useful against undesired LAB (such as histamine producers or spoiler) in vinification, and also for proper control of the wine malolactic fermentation (Yurdugül and Bozoglu 2002).

References

Abriouel H, Lucas R, Ben Omar N et al (2010) Potential applications of the cyclic peptide enterocin AS-48 in the preservation of vegetable foods and beverages. Probiot Antimicrob Prot 2:77–89

Allende A, Martínez B, Selma V et al (2007) Growth and bacteriocin production by lactic acid bacteria in vegetable broth and their effectiveness at reducing *Listeria monocytogenes* in vitro and in fresh-cut lettuce. Food Microbiol 24:759–766

Ananou S, Gálvez A, Martínez-Bueno M et al (2005) Synergistic effect of enterocin AS-48 in combination with outer membrane permeabilizing treatments against *Escherichia coli* O157:H7. J Appl Microbiol 99:1364–1372

Atrih A, Rekhif N et al (1993) Detection and characterization of a bacteriocin produced by *Lactobacillus plantarum* C19. Can J Microbiol 39:1173–1179

Bari ML, Ukuku DO, Kawasaki T et al (2005) Combined efficacy of nisin and pediocin with sodium lactate, citric acid, phytic acid, and potassium sorbate and EDTA in reducing the *Listeria monocytogenes* population of inoculated fresh-cut produce. J Food Prot 68:1381–1387

Bartowsky EJ (2009) Bacterial spoilage of wine and approaches to minimize it. Lett Appl Microbiol 48:149–156

Basanta A, Herranz C, Gutiérrez J et al (2009) Development of bacteriocinogenic strains of *Saccharomyces cerevisiae* heterologously expressing and secreting the leaderless enterocin

L50 peptides L50A and L50B from *Enterococcus faecium* L50. Appl Environ Microbiol 75:2382–2392

Basanta A, Sánchez J, Gómez-Sala B et al (2008) Antimicrobial activity of *Enterococcus faecium* L50, a strain producing enterocins L50 (L50A and L50B), P and Q, against beer-spoilage lactic acid bacteria in broth, wort (hopped and unhopped), and alcoholic and non-alcoholic lager beers. Int J Food Microbiol 125:293–307

Bauer R, Nel HA, Dicks LMT (2003) Pediocin PD-1 as a method to control growth of *Oenococcus oeni* in wine. Am J Enol Vitic 54:86–91

Ben Omar N, Abriouel H, Keleke S et al (2008) Bacteriocin producing *Lactobacillus* strains isolated from poto poto, a Congolese fermented maize product, and genetic fingerprinting of their plantaricin operons. Int J Food Microbiol 127:18–25

Ben Omar N, Abriouel H, Lucas R et al (2006) Isolation of bacteriocinogenic *Lactobacillus plantarum* strains from ben saalga, a traditional fermented gruel from Burkina Faso. Int J Food Microbiol 112:44–50

Bennik MHJ, Van Overbeek W, Smid EJ et al (1999) Biopreservation in modified atmosphere stored mungbean sprouts: the use of vegetable-associated bacteriocinogenic lactic acid bacteria to control the growth of *Listeria monocytogenes*. Lett Appl Microbiol 28:226–232

Beuchat LR (1997) Traditional fermented foods. In: Doyle MP, Beuchat LR, Montville TJ (eds) Food microbiology—fundamentals and frontiers. ASM, Washington, DC, pp 629–648

Breidt F, Crowley KA, Fleming HP (1995) Controlling cabbage fermentations with nisin and nisin-resistant *Leuconostoc mesenteroides*. Food Microbiol 12:109–116

Brown AL, Brooks JC, Karunasena E et al (2011) Inhibition of *Escherichia coli* O157:H7 and *Clostridium sporogenes* in spinach packaged in modified atmospheres after treatment combined with chlorine and lactic acid bacteria. J Food Sci 76:M427–M432

Cabo ML, Pastoriza L, Sampedro G et al (2001) Joint effect of nisin, CO_2, and EDTA on the survival of *Pseudomonas aeruginosa* and *Enterococcus faecium* in a food model system. J Food Prot 64:1943–1948

Cabo ML, Torres B, Herrera JJ et al (2009) Application of nisin and pediocin against resistance and germination of *Bacillus* spores in sous vide products. J Food Prot 72:515–623

Cálix-Lara TF, Rajendran M, Talcott ST et al (2014) Inhibition of *Escherichia coli* O157:H7 and *Salmonella enterica* on spinach and identification of antimicrobial substances produced by a commercial Lactic Acid Bacteria food safety intervention. Food Microbiol 38:192–200

Chang JY, Chang HC (2011) Growth inhibition of foodborne pathogens by kimchi prepared with bacteriocin-producing starter culture. J Food Sci 76:M72–M78

Chihib NE, Crepin T, Delattre G et al (1999a) Involvement of cell envelope in nisin resistance of *Pectinatus frisingensis*, a Gram-negative, strictly anaerobic beer-spoilage bacterium naturally sensitive to nisin. FEMS Microbiol Lett 177:167–175

Chihib NE, Monnerat L, Membré JM et al (1999b) Nisin, temperature and pH effects on growth and viability of *Pectinatus frisingensis*, a Gram-negative, strictly anaerobic beer-spoilage bacterium. J Appl Microbiol 87:438–446

Choi MH, Park YH (2000) Selective control of lactobacilli in kimchi with nisin. Lett Appl Microbiol 30:173–177

Choi SY, Beuchat LR (1994) Growth inhibition of *Listeria monocytogenes* by a bacteriocin of *Pediococcus acidilactici* M during fermentation of kimchi. Food Microbiol 11:301–307

Cobo Molinos A, Abriouel H, Ben Omar N et al (2005) Effect of immersion solutions containing enterocin AS-48 on *Listeria monocytogenes* in vegetable foods. Appl Environ Microbiol 71:7781–7787

Cobo Molinos A, Abriouel H, Lucas R et al (2008a) Inhibition of *Bacillus cereus* and *Bacillus weihenstephanensis* in raw vegetables by application of washing solutions containing enterocin AS-48 alone and in combination with other antimicrobials. Food Microbiol 25:762–770

Cobo Molinos A, Abriouel H, Lucas López R et al (2008b) Combined physico-chemical treatments based on enterocin AS-48 for inactivation of Gram-negative bacteria in soybean sprouts. Food Chem Toxicol 46:2912–2921

Cobo Molinos A, Abriouel H, Ben Omar N et al. (2008c) Inactivation of *Listeria monocytogenes* in raw fruits by enterocin AS-48. J. Food Prot 71:2460–2467

Cobo Molinos A, Abriouel H, Lucas López R et al (2009a) Enhanced bactericidal activity of enterocin AS-48 in combination with essential oils, natural bioactive compounds and chemical preservatives against Listeria monocytogenes in ready-to-eat salad. Food Chem Toxicol 47:2216–2223

Cobo Molinos A, Lucas R, Abriouel H et al (2009b) Inhibition of *Salmonella enterica* cells in deli-type salad by enterocin AS-48 in combination with other antimicrobials. Probiot Antimicrob Prot 1:85–90

Corsetti A, Settanni L, Van Sinderen D (2004) Characterization of bacteriocin-like inhibitory substances (BLIS) from sourdough lactic acid bacteria and evaluation of their in vitro and in situ activity. J Appl Microbiol 96:521–534

Corsetti A, Settanni L (2007) Lactobacilli in sourdough fermentation. Food Res Int 40:539–558

Daeschel MA, McKeney MC, McDonald LC (1990) Bacteriocidal activity of *Lactobacillus plantarum* C-11. Food Microbiol 7:91–98

Dalié DKD, Deschamps AM, Richard-Forget F (2010) Lactic acid bacteria – potential for control of mould growth and mycotoxins: a review. Food Control 21:370–380

De Vuyst L, Vrancken G, Ravyts F et al (2009) Biodiversity, ecological determinants, and metabolic exploitation of sourdough microbiota. Food Microbiol 26:666–675

Delgado A, Brito D, Peres C et al (2005) Bacteriocin production by *Lactobacillus pentosus* B96 can be expressed as a function of temperature and NaCl concentration. Food Microbiol 22:521–528

Delves-Broughton J, Blackburn P, Evans RJ et al (1996) Applications of the bacteriocin, nisin. Anton Van Leeuw 69:193–202

Diels AM, De Taeye J, Michiels CW (2004) High-pressure homogenisation sensitises Escherichia coli to lysozyme and nisin. Commun Agric Appl Biol Sci 69:115–117

Diels AM, De Taeye J, Michiels CW (2005) Sensitisation of Escherichia coli to antibacterial peptides and enzymes by high-pressure homogenization. Int J Food Microbiol 105:165–175

Díez L, Rojo-Bezares B, Zarazaga M et al (2012) Antimicrobial activity of pediocin PA-1 against *Oenococcus oeni* and other wine bacteria. Food Microbiol 31:167–172

Du Toit M, Pretorius IS (2000) Microbial spoilage and preservation of wine: using weapons from nature's own arsenal. South Afr J Enol Vit 21:74–96

Franz CMAP, Schillinger U, Holzapfel WH (1996) Production and characterization of enterocin 900, a bacteriocin produced by *Enterococcus faecium* BFE 900 from black olives. Int J Food Microbiol 29:255–270

Galvez A, Lopez RL, Abriouel H et al (2008) Application of bacteriocins in the controlof foodborne pathogenic and spoilage bacteria. Crit Rev Biotechnol 28:125–152

Gänzle MG, Vogel RF (2003) Contribution of reutericyclin production to the stable persistence of *Lactobacillus reuteri* in an industrial sourdough fermentation. Int J Food Microbiol 80:31–45

Gao Y, Jia S, Gao Q et al (2010) A novel bacteriocin with a broad inhibitory spectrum produced by *Lactobacillus sake* C2, isolated from traditional Chinese fermented cabbage. Food Control 21:76–81

García-Ruiz A, Teresa Requena T, Peláez C et al (2013) Antimicrobial activity of lacticin 3147 against oenological lactic acid bacteria. Combined effect with other antimicrobial agents. Food Control 32:477–483

Ge J, Ping W, Song G et al (2009) Paracin 1.7, a bacteriocin produced by *Lactobacillus paracasei* HD1.7 isolated from Chinese cabbage sauerkraut, a traditional Chinese fermented vegetable food. Wei Sheng Wu Xue Bao 49:609–616

Gilbreth SE, Somkuti GA (2005) Thermophilin 110: a bacteriocin of *Streptococcus thermophilus* ST110. Curr Microbiol 51:175–182

Grande MJ, Abriouel H, Lucas R et al (2007a) Efficacy of enterocin AS-48 against bacilli in ready-to-eat vegetable soups and purees. J Food Prot 70:2339–2345

Grande MJ, López RL, Abriouel H et al (2007b) Treatment of vegetable sauces with enterocin AS-48 alone or in combination with phenolic compounds to inhibit proliferation of *Staphylococcus aureus*. J Food Prot 70:405–411

Grande MJ, Lucas R, Abriouel H et al (2005) Control of *Alicyclobacillus acidoterrestris* in fruit juices by enterocin AS-48. Int J Food Microbiol 104:289–297

Grande MJ, Lucas R, Abriouel H et al (2006a) Inhibition of toxicogenic *Bacillus cereus* in rice-based foods by enterocin AS-48. Int J Food Microbiol 106:185–194

Grande MJ, Lucas R, Abriouel H et al (2006b) Inhibition of *Bacillus licheniformis* LMG 19409 from ropy cider by enterocin AS-48. J Appl Microbiol 101:422–428

Gutiérrez J, Criado R, Martín M et al (2005) Production of enterocin P, an antilisterial pediocin-like bacteriocin from *Enterococcus faecium* P13, in *Pichia pastoris*. Antimicrob Agents Chemother 49:3004–3008

Harris LJ, Fleming HP, Klaenhammer TR (1992a) Novel paired starter culture system for sauerkraut, consisting of a nisin-resistant *Leuconostoc mesenteroides* strain and a nisin-producing *Lactococcus lactis* strain. Appl Environ Microbiol 58:1484–1489

Harris LJ, Fleming HP, Klaenhammer TR (1992b) Characterization of two nisin-producing *Lactococcus lactis* subsp. *Lactis* strains isolated from a commercial sauerkraut fermentation. Appl Environ Microbiol 58:1477–1483

Hata T, Alemu M, Kobayashi M et al (2009) Characterization of a bacteriocin produced by *Enterococcus faecalis* N1-33 and its application as a food preservative. J Food Prot 72:524–530

Huang Y, Luo Y, Zhai Z et al (2009) Characterization and application of an anti-*Listeria* bacteriocin produced by *Pediococcus pentosaceus* 05-10 isolated from Sichuan Pickle, a traditionally fermented vegetable product from China. Food Contr 20:1030–1035

Iu J, Mittal GS, Griffiths MW (2001) Reduction in levels of *Escherichia coli* O157:H7 in apple cider by pulsed electric fields. J of Food Prot 64:964–969

Jay JM, Loessner MJ, Golden A (2005) Modern food microbiology. Aspen Publishers Inc., Gaithersburg, MD, USA

Jespersen L, Jakobsen M (1996) Specific spoilage organisms in breweries and laboratory media for their detection. Int J Food Microbiol 33:139–155

Jimenez-Diaz R, Rios-Sanchez RM, Desmazeaud M et al (1993) Plantaricins S and T, two new bacteriocins produced by *Lactobacillus plantarum* LPCO10 isolated from a green olive fermentation. Appl Environ Microbiol 59:1916–1924

Jiménez-Díaz R, Rios-Sánchez RM, Desmazeaud M et al (1993) Plantaricins S and T, two new bacteriocins produced by *Lactobacillus plantarum* LPCO10 isolated from a green olive fermentation. Appl Environ Microbiol 59:1416–1424

Kato T, Inuzuka L, Kondo M et al (2001) Growth of nisin-producing lactococci in cooked rice supplemented with soybean extract and its application to inhibition of *Bacillus subtilis* in rice miso. Biosci Biotechnol Biochem 65:330–337

Knoll C, Divol B, du Toit M (2008) Genetic screening of lactic acid bacteria of oenological origin for bacteriocin-encoding genes. Food Microbiol 25:983–991

Komitopoulou E, Boziaris IS, Davies EA et al (1999) *Alicyclobacillus acidoterrestris* in fruit juices and its control by nisin. Int J Food Sci Technol 34:81–85

Kostinek M, Specht I, Edward VA et al (2007) Characterisation and biochemical properties of predominant lactic acid bacteria from fermenting cassava for selection as starter cultures. Int J Food Microbiol 114:342–351

Kusano K, Yamada H, Niwa M, Yamasoto K (1997) *Propionibacterium cyclohexanicum* sp. nov. a new-tolerant x-cyclohexyl fatty acid-containing propionibacterium isolated from spoiled orange juice. Int J Syst Bacteriol 47:825–831

Leroy F, De Winter T, Moreno MRF et al (2007) The bacteriocin producer *Lactobacillus amylovorus* DCE 471 is a competitive starter culture for type II sourdough fermentations. J Sci Food Agric 87:1726–1736

Leverentz B, Conway WS, Camp MJ et al (2003) Biocontrol of *Listeria monocytogenes* on fresh-cut produce by treatment with lytic bacteriophages and a bacteriocin. Appl Environ Microbiol 69:4519–4526

Liang Z, Cheng Z, Mittal GS (2006) Inactivation of spoilage microorganisms in apple cider using a continuous flow pulsed electric field system. LWT Food Sci Technol 39:351–357

Liang Z, Mittal GS, Griffiths MW (2002) Inactivation of *Salmonella* Typhimurium in orange juice containing antimicrobial agents by pulsed electric field. J Food Prot 65:1081–1087

Lonvaud-Funel A, Joyeux A (1993) Antagonism between lactic acid bacteria of wines: inhibition of *Leuconostoc oenos* by *Lactobacillus plantarum* and *Pediococcus pentosaceus*. Food Microbiol 10:411–419

Lucas R, Grande MJ, Abriouel H et al (2006) Application of the broad-spectrum bacteriocin enterocin AS-48 to inhibit *Bacillus coagulans* in low-pH canned fruit and vegetable foods. Food Chem Toxicol 44:1774–1781

Lynch MF, Tauxe RV, Hedberg CW (2009) The growing burden of foodborne outbreaks due to contaminated fresh produce: risks and opportunities. Epidemiol Infect 137:307–315

Maldonado A, Ruiz-Barba JL, Floriano B et al (2002) The locus responsible for production of plantaricin S, a class IIb bacteriocin produced by *Lactobacillus plantarum* LPCO10, is widely distributed among wild-type *Lact. plantarum* strains isolated from olive fermentations. Int J Food Microbiol 77:117–124

Martín-Belloso O, Elez-Martínez P (2005) Food safety aspects of pulsed electric fields. In: Sun DW (ed) Emerging technologies for food processing. Elsevier, Amsterdam, pp 183–217

Martínez-Viedma P, Abriouel H, Ben Omar N et al (2008a) Inactivation of exopolysaccharide and 3-hydroxypropionaldehyde-producing lactic acid bacteria in apple juice and apple cider by enterocin AS-48. Food Chem Toxicol 46:1143–1151

Martínez-Viedma P, Sobrino A, Ben Omar N et al (2008b) Enhanced bactericidal effect of High-Intensity Pulsed-Electric Field treatment in combination with enterocin AS-48 against *Salmonella enterica* in apple juice. Int J Food Microbiol 128:244–249

Martínez Viedma P, Abriouel H, Ben Omar N (2009a) Anti-staphylococcal effect of enterocin AS-48 in bakery ingredients of vegetable origin, alone and in combination with selected anti-microbials. J Food Sci 74:M384–M389

Martínez-Viedma P, Abriouel H, Ben Omar N et al (2009b) Assay of enterocin AS-48 for inhibition of foodborne pathogens in desserts. J Food Prot 72:1654–1659

Martínez-Viedma P, Abriouel H, Sobrino A et al (2009c) Effect of enterocin AS-48 in combination with high-intensity pulsed-electric field treatment against the spoilage bacterium *Lactobacillus diolivorans* in apple juice. Food Microbiol 26:491–496

Martínez Viedma P, Abriouel H, Ben Omar N et al (2010a) Effect of enterocin EJ97 against *Geobacillus stearothermophilus* vegetative cells and endospores in canned foods and beverages. Eur Food Res Technol 230:513–519

Martínez Viedma P, Ercolini D, Ferrocino I et al (2010b) Effect of polythene film activated with enterocin EJ97 in combination with EDTA against *Bacillus coagulans*. LWT Food Sci Technol 43:514–518

Menteş Ö, Ercan R, Akçelik M (2007) Inhibitor activities of two *Lactobacillus* strains, isolated from sourdough, against rope-forming *Bacillus* strains. Food Control 18:359–363

Messens W, De Vuyst L (2002) Inhibitory substances produced by lactobacilli isolated from sourdoughs – a review. Int J Food Microbiol 72:31–43

Messens W, Neysens P, Vansieleghem W et al (2002) Modeling growth and bacteriocin production by Lactobacillus amylovorus DCE 471 in response to temperature and pH values used for sourdough fermentations. Appl Environ Microbiol 68:1431–1435

Mittal GS, Griffiths MW (2005) Pulsed electric field processing of liquid foods and beverages. In: Sun DW (ed) Emerging technologies for food processing. Elsevier, Amsterdam, pp 99–139

Mokhtari A, Moore CM, Yang H et al (2006) Consumer-phase *Salmonella enterica* serovar Enteritidis risk assessment for egg-containing food products. Risk Anal 26:753–768

Mosqueda-Melgar J, Elez-Martínez P, Raybaudi-Massilia RM et al (2008) Effects of pulsed electric fields on pathogenic microorganisms of major concern in fluid foods: a review. Crit Rev Food Sci Nutr 48:747–759

Narbutaite V, Fernandez A, Horn N et al (2007) Influence of baking enzymes on antimicrobial activity of five bacteriocin-like inhibitory substances produced by lactic acid bacteria isolated from Lithuanian sourdoughs. Lett Appl Microbiol 47:555–560

Navarro L, Zarazaga M, Sáenz J et al (2000) Bacteriocin production by lactic acid bacteria isolated from Rioja red wines. J Appl Microbiol 88:44–51

Nel HA, Bauer R, Wolfaardt GM et al (2002) Effect of bacteriocins pediocin PD-1, plantaricin 423 and nisin on biofilms of *Oenococcus oeni* on a stainless steel surface. Am J Enol Vitic 53:191–196

Nguyen P, Mittal GS (2007) Inactivation of naturally occurring microorganisms in tomato juice using pulsed electric field (PEF) with and without antimicrobials. Chem Eng Proc 46:360–365

Ogden K, Tubb RS (1985) Inhibition of beer-spoilage lactic acid bacteria by nisin. J Inst Brew 91:390–392

Ogden K, Waites MJ (1986) The action of nisin on beer spoilage lactic acid bacteria. J Inst Brew 92:463–467

Pathanibul P, Taylor TM, Davidson PM et al (2009) Inactivation of *Escherichia coli* and *Listeria innocua* in apple and carrot juices using high pressure homogenization and nisin. Int J Food Microbiol 129:316–320

Peña WE, de Massaguer PR, Teixeira LQ (2009) Microbial modeling of thermal resistance of *Alicyclobacillus acidoterrestris* CRA7152 spores in concentrated orange juice with nisin addition. Braz J Microbiol 40:601–611

Pepe O, Blaiotta G, Moschetti G et al (2003) Rope-producing strains of *Bacillus* spp. from wheat bread and strategy for their control by lactic acid bacteria. Appl Environ Microbiol 69:2321–2329

Pérez Pulido R, Ben Omar N, Abriouel H et al (2005) Microbiological study of lactic acid fermentation of caper berries by molecular and culture-dependent methods. Appl Environ Microbiol 71:7872–7879

Radler F (1990) Possible use of nisin in wine-making. II Experiments to control lactic acid bacteria in the production of wine. Am J Enol Vit 41:7–11

Randazzo CL, Pitino I, Scifo GO et al (2009) Biopreservation of minimally processed iceberg lettuces using a bacteriocin produced by *Lactococcus lactis* wild strain. Food Control 20:756–763

Rojo-Bezares B, Sáenz Y, Zarazaga M et al (2007) Antimicrobial activity of nisin against *Oenococcus oeni* and other wine bacteria. Int J Food Microbiol 116:32–36

Rosenquist H, Hansen A (1998) The antimicrobial effect of organic acids, sour dough and nisin against *Bacillus subtilis* and *B. licheniformis* isolated from wheat bread. J Appl Microbiol 85:621–631

Ruiz-Barba JL, Cathcart DP, Warner PJ et al (1994) Use of *Lactobacillus plantarum* LPCO10, a bacteriocin producer, as a starter culture of Spanish-style green olive fermentations. Appl Environ Microbiol 60:2059–2064

Sánchez J, Borrero J, Gómez-Sala B et al (2008) Cloning and heterologous production of Hiracin JM79, a Sec-dependent bacteriocin produced by *Enterococcus hirae* DCH5, in lactic acid bacteria and *Pichia pastoris*. Appl Environ Microbiol 74:2471–2479

Sánchez Valenzuela A, Díaz Ruiz G, Ben Omar N et al (2008) Inhibition of food poisoning and pathogenic bacteria by *Lactobacillus plantarum* strain 2.9 isolated from ben saalga, both in a culture medium and in food. Food Control 19:842–848

Schillinger U, Becker B, Vignolo G et al (2001) Efficacy of nisin in combination with protective cultures against *Listeria monocytogenes* Scott A in tofu. Int J Food Microbiol 71:159–168

Schoeman H, Vivier MA, du Toit M et al (1999) The development of bactericidal yeast strains by expressing the *Pediococcus acidilactici* pediocin gene (*pedA*) in *Saccharomyces cerevisiae*. Yeast 15:647–656

Seseña S, Sánchez I, Palop L (2004) Genetic diversity (RAPD-PCR) of lactobacilli isolated from "Almagro" eggplant fermentations from two seasons. FEMS Microbiol Lett 238:159–165

Settanni L, Massitti O, Van Sinderen D et al (2005) In situ activity of a bacteriocin-producing *Lactococcus lactis* strain. Influence on the interactions between lactic acid bacteria during sourdough fermentation. J Appl Microbiol 99:670–681

Settanni L, Corsetti A (2008) Application of bacteriocins in vegetable food biopreservation. Int J Food Microbiol 121:123–138

Silveira AC, Conesa A, Aguayo E et al (2008) Alternative sanitizers to chlorine for use on fresh-cut "Galia" (*Cucumis melo* var. *catalupensis*) melon. J Food Sci 73:405–411

Strasser de Saad AM, Manca de Nadra MC (1993) Characterization of bacteriocin produced by *Pediococcus pentosaceus* from wine. J Appl Bacteriol 74:406–410

Strasser de Saad AM, Pasteris SE, Manca de Nadra MC (1995) Production and stability of pediocin N5p in grape juice medium. J Appl Bacteriol 78:473–476

Tamang JP, Tamang B, Schillinger U et al (2009) Functional properties of lactic acid bacteria isolated from ethnic fermented vegetables of the Himalayas. Int J Food Microbiol 135:28–33

Thomas LV, Clarkson MR, Delves-Broughton J (2000) Nisin. In: Naidu AS (ed) Natural food antimicrobial systems. CRC, Boca Raton, FL, pp 463–524

Thomas LV, Ingram RE, Bevis HE et al (2002) Effective use of nisin to control *Bacillus* and *Clostridium* spoilage of a pasteurized mashed potato product. J Food Prot 65:1580–1585

Todorov SD, Dicks LMT (2005) Characterization of bacteriocins produced by lactic acid bacteria isolated from spoiled black olives. J Basic Microbiol 45:312–322

Torriani S, Orsi C, Vescovo M (1997) Potential of *Lactobacillus casei*, culture permeate, and lactic acid to control microorganisms in ready-to-use vegetables. J Food Prote 60:1564–1567

Trias R, Badosa E, Montesinos E et al (2008a) Bioprotective *Leuconostoc* strains against *Listeria monocytogenes* in fresh fruits and vegetables. Int J Food Microbiol 127:91–98

Trias R, Bañeras L, Badosa E et al (2008b) Bioprotection of Golden Delicious apples and Iceberg lettuce against foodborne bacterial pathogens by lactic acid bacteria. Int J Food Microbiol 123:50–60

Uhlman L, Schillinger U, Rupnow JR et al (1992) Identification and characterization of two bacteriocin-producing strains of *Lactococcus lactis* isolated from vegetables. Int J Food Microbiol 16:141–151

Ukuku DO, Bari ML, Kawamoto S et al (2005) Use of hydrogen peroxide in combination with nisin, sodium lactate and citric acid for reducing transfer of bacterial pathogens from whole melon surfaces to fresh-cut pieces. Int J Food Microbiol 104:225–233

Ukuku DO, Fett WF (2004) Effect of nisin in combination with EDTA, sodium lactate, and potassium sorbate for reducing *Salmonella* on whole and fresh-cut cantaloupes. J Food Prot 67:2143–2150

Ukuku DO, Zhang H, Huang L (2009) Growth parameters of *Escherichia coli* O157:H7, *Salmonella* spp., *Listeria monocytogenes*, and aerobic mesophilic bacteria of apple cider amended with nisin-EDTA. Foodb Pathog Dis 6:487–494

Unicomb L, Bird P, Dalton C (2003) Outbreak of *Salmonella* Potsdam associated with salad dressing at a restaurant. Commun Dis Intell 27:508–512

Valerio F, De Bellis P, Lonigro SL et al (2008) Use of *Lactobacillus plantarum* fermentation products in bread-making to prevent *Bacillus subtilis* ropy spoilage. Int J Food Microbiol 122:328–332

Valerio F, Favilla M, De Bellis P et al (2009) Antifungal activity of strains of lactic acid bacteria isolated from a semolina ecosystem against *Penicillium roqueforti*, *Aspergillus niger* and *Endomyces fibuliger* contaminating bakery products. Syst Appl Microbiol 32:438–448

Van Reenen CA, Chikindas ML, van Zyl WH et al (2002) Characterisation and heterologous expression of a class IIa bacteriocin, plantaricin 423 from *Lactobacillus plantarum* 423, in *Saccharomyces cerevisiae*. Int J Food Microbiol 81:29–40

Vaughan A, Eijsink VJ, O'Sullivan TF et al (2001) An analysis of bacteriocins produced by lactic acid bacteria isolated from malted barley. J Appl Microbiol 91:131–138

Vaughan A, O'Sullivan T, van Sinderen D (2005) Enhancing the microbiological stability of malt and beer - a review. J Inst Brewing 111:355–371

Vega Leal-Sánchez M, Ruiz-Barba JL, Sánchez AH et al (2003) Fermentation profile and optimization of green olive fermentation using *Lactobacillus plantarum* LPCO10 as a starter culture. Food Microbiol 20:421–430

Walker M, Phillips CA (2008) The effect of preservatives on *Alicyclobacillus acidoterrestris* and *Propionibacterium cyclohexanicum* in fruit juice. Food Control 19:974–981

Wu Y, Mittal GS, Griffiths MW (2005) Effect of Pulsed Electric Field on the inactivation of micro-organisms in grape juices with and without antimicrobials. Biosyst Eng 90:1–7

Yamazaki K, Mukarami M, Kawai Y et al (2000) Use of nisin for inhibition of *Alicyclobacillus acidoterrestris* in acidic drinks. Food Microbiol 17:315–320

Yanagida F, Srionnual S, Chen YS (2008) Isolation and characteristics of lactic acid bacteria from koshu vineyards in Japan. Lett Appl Microbiol 47:139–320

Yoon MY, Kim YJ, Hwang HJ (2008) Properties and safety aspects of *Enterococcus faecium* strains isolated from *Chungkukjang*, a fermented soy product. LWT Food Sci Technol 41:925–933

Yurdugül S, Bozoglu F (2002) Studies on an inhibitor produced by lactic acid bacteria of wines on the control of malolactic fermentation. Eur Food Res Technol 215:38–41

Yuste J, Fung DY (2004) Inactivation of *Salmonella* Typhimurium and *Escherichia coli* O157:H7 in apple juice by a combination of nisin and cinnamon. J Food Prot 67:371–377

Zhao L, Wang Y, Wang S et al (2014) Inactivation of naturally occurring microbiota in cucumber juice by pressure treatment. Int J Food Microbiol 174:12–18

Chapter 9
Regulations

9.1 The Challenge of Taking Bioprotection Strategies from the Lab to the Market

Application of bioprotection strategies in food preservation may be restricted by laws from different countries that may differ considerably in their fundamentals end practical effects. Some of them are related directly to addition of bioprotectants to foods, but others not less important may be related to apparently secondary aspects such as labelling, packaging, export, use of biological agents or genetically modified organisms.

Licensing antimicrobial preparations as food preservatives is perhaps the most complicated way to follow. Nisin is the only bacteriocin currently approved as a food preservative (E234). Nisin was assessed to be safe for food use by the Joint Food and Agriculture Organization/World Health Organization Expert Committee on Food Additives in 1969, and was added to the European food additive list in 1983 (Directive 83/463/EEC; Directive 95/2/EC); (European Economic Community 1983; European Parliament and Council 1995). It was approved in 1988 by the US Food and Drug Agency (FDA) for use in pasteurized processed cheese spreads. The initial approval was followed by other licensed applications (e.g., FSIS 2002). Nisin is legally used in over 80 countries (Adams 2003). However, there are major differences in national legislations concerning the presence and levels of nisin in food products.

Nisin (in the commercial forms Nisaplin™ and Chrisin™) is a lyophilized product obtained from a microbial fermentation. The industrially fermented products are regulated under general food laws. Concentrates or lyophilised powders obtained from fermentates may be added to foods as ingredients. Many commercial preparations currently on the market are sold as ingredients, or shelf-life extenders. Alta™ 2351 and Fargo 23 are natural food ingredients with antilisterial activity produced by bacteriocinogenic strains through a fermentation process. Alta™ 2351 is labelled as "cultured dairy solids (Skim Milk, Dextrose, Whey, and Lactic Acid Culture)."

© The Author(s) 2014
A. Gálvez et al., *Food Biopreservation*, SpringerBriefs in Food, Health, and Nutrition, DOI 10.1007/978-1-4939-2029-7_9

It is "an ingredient with functionality against outgrowth of *Listeria* in dairy based products, and small spectrum lactic acid bacteria inhibition." Such products are approved in the US and commercialized in several other countries as well as shelf life extenders for application in a variety of food products.

According to Directive 2000/13/EC (European Parliament and Council 2000) on food labelling, "'ingredient' shall mean any substance, including additives, used in the manufacture or preparation of a foodstuff and still present in the finished product, even if in altered form." Fermented milk or whey concentrates or lyophilised preparations (regardless of whether they contain or not bacteriocins) can be added as ingredients in the preparation of dairy foods. In the European Union, approval of bacteriocins for application as food additives or preservatives must comply with specifications given in Regulation 1333/2008/EC (European Parliament and Council (2008b), which harmonises the use of food additives in foods in the European Community and updates Directive 89/107/EEC (European Commission 1988) concerning food additives authorized for use in foodstuffs intended for human consumption and Directive 95/2/EC (European Parliament and Council 1995) on food additives other than colours and sweeteners, and Regulation 258/97/EC (European Parliament and Council 1997) on novel foods and novel ingredients. Food additives are substances that are not normally consumed as food itself but are added to food intentionally for a technological purpose described in the above Regulation, such as the preservation of food. Preservatives are considered a functional class of food additives: 'preservatives' are substances which prolong the shelf-life of foods by protecting them against deterioration caused by micro-organisms and/or which protect against growth of pathogenic micro-organisms. Approved food additives must be listed in the Community lists and shall specify: (a) the name of the food additive and its E number; (b) the foods to which the food additive may be added; (c) the conditions under which the food additive may be used; (d) if appropriate, whether there are any restrictions on the sale of the food additive directly to the final consumer. Approved specifications should include information to adequately identify the food additive, including origin, and to describe the acceptable criteria of purity. Added preservatives must be listed in food labels with their specific name or EC number (Directive 2000/13/EC; European Parliament and Council 2000). The use and maximum levels of a food additive should take into account the intake of the food additive from other sources and the exposure to the food additive by special groups of consumers (e.g. allergic consumers). The risk assessment and approval of food additives should be carried out in accordance with the procedure laid down in Regulation (EC) No 1331/2008 (European Parliament and Council 2008a) establishing a common authorisation procedure for food additives, food enzymes and food flavourings. Food additives which were permitted before 20 January 2009 shall be subject to a new risk assessment carried out by the Authority.

Bacteriocin preparations could also be applied as processing aids in food manufacture. Directive 2000/13/EC (European Parliament and Council 2000) and Regulation 1333/2008/EC (European Parliament and Council 2008b) do not cover processing aids, but according to Regulation 1333/2008/EC 'processing aid' shall mean any substance which: (1) is not consumed as a food by itself; (2) is intentionally used in the processing of raw materials, foods or their ingredients, to fulfill a certain technological purpose during treatment or processing; and (3) may result in the unintentional but technically unavoidable presence in the final product of residues of the substance or its derivatives provided they do not present any health risk and do not have any technological effect on the final product. According to this definition, bacteriocins could be applied as processing aids for the preservation of food ingredients, whereby the bacteriocin has no preservative or technological effect in the final food product.

Application of bacteriocins in activated packagings must follow specifications of Directive 2002/72/EC (European Parliament and Council 2002b) concerning plastic materials and articles intended to come into contact with foodstuffs and Regulation (EC) No 1935/2004 (European Parliament and Council 2004) on materials and articles intended to come into contact with food: Active food contact materials are designed to deliberately incorporate 'active' components intended to be released into the food or to absorb substances from the food: "'active food contact materials and articles' (hereinafter referred to as active materials and articles) means materials and articles that are intended to extend the shelf-life or to maintain or improve the condition of packaged food. They are designed to deliberately incorporate components that would release or absorb substances into or from the packaged food or the environment surrounding the food." Substances deliberately incorporated into active materials and articles to be released into the food or the environment surrounding the food shall be authorised and used in accordance with the relevant Community provisions applicable to food, and shall comply with the provisions of this Regulation and its implementing measures. These substances shall be considered as ingredients. Covering or coating materials forming part of the food and possibly being consumed with it (such as edible coatings) do not fall within the scope of this Regulation.

Bacteriocin-producing strains may be applied as starter or bioprotective cultures with the aim of contributing to microbiological safety (Aymerich et al. 2008). For example, Bactoferm F-Lc (Christian Hansen, Denmark) is an antilisterial mixed culture of *Pediococcus acidilactici* and *Lactobacillus curvatus* producing pediocin and sakacin A, respectively for application in fermented sausages. The same company also sells bioprotective cultures containing *Lactobacillus sakei*, and *Leuconostoc carnosum* 4010 for meat products packed under vacuum or modified atmosphere packaging (MAP), and a nisin-producing *Lactococcus lactis* preparation. Danisco (Copenhaguen, Denmark) markets a series of protective cultures (HOLDBAC™) for specific applications in meat and dairy foods based on their capacity to produce bcteriocins as well as other antimicrobial compounds and their competition in food systems. Such preparations include mainly strains of *Lactobacillus plantarum*, *Lactobacillus rhamnosus*, *L. sakei*, *Lactobacillus paracasei* and *Propionibacterium freundenreichii* subsp. *shermanii*), whose primary functionalities are growth control of Gram-positive pathogens such as *Listeria*, spoilage microorganisms such as yeasts and moulds, heterofermentative lactic bacteria, and enterococci. Such strains have not been subjected to genetic modification, but the company advertises that local regulations should always be consulted concerning the status of these products as legislation regarding their use in food may vary from country to country. Similar commercial preparations consisting of bioactive LAb cultures are also available in the US, like for example LactiGuard™, under approval of the US-FDA and the USDA-FSIS.

From a regulatory point of view, bacteriocin-producing strains fall in the category of microbial cultures. In the United States, a new strain of micro-organism for use in food can either be classified as an additive or as a Generally Recognised as Safe (GRAS) substance (Wessels et al. 2004). Food additives are defined in a broad sense as "anything that might come into contact with food (excluding GRAS substances)," and require pre-market approval by the US FDA based on toxicological and efficacy data. The consideration of GRAS status is based on the availability of enough information relevant to the substance safe use for a given intended purpose: "generally recognized, among experts qualified by scientific training and experience to evaluate its safety, as having been adequately shown through scientific procedures (or, in the case as a substance used in food prior to January 1, 1958, through either scientific procedures or experience based on common use in food) to be safe under the conditions of its intended use" (US Food and Drug Administration 1999). The intended use is an essential part of the GRAS status concept. For instance, the GRAS status of a given strain for use in a yogurt product is not valid for the same strain in infant formulae (Wessels et al. 2004). The GRAS status is determined by qualified experts, not by the FDA. The food company that uses the bacterium assumes complete responsibility, regardless of its GRAS status.

Within the European Union, microbial food cultures with a long history of safe use are considered as traditional food ingredients, and covered by general European food law (Regulation 178/2002/EC; European Parliament and Council 2002a). Microbial cultures must also be safe for their intended use. Novel use of microbial cultures is regulated in Regulation 258/97/EC (European Parliament and Council 1997) if a microorganism has not been consumed in a significant degree before May 15, 1997. This may apply to selected bacteriocin producer strains isolated from a source different than the food where they will be applied. There is also an ongoing dispute in Europe regarding the food category of starter cultures with protective properties, since they may be considered as cultures with specific technological effects (preservatives). This may contradict current regulations on approval of new preservatives, given the long history of consumption of fermented foods and the fact that the original and primary purpose of fermenting food was to achieve a preservation effect. In this respect, guidance documents from the European Food Safety Authority (EFSA 2007, 2008) established a pre-market safety assessment of selected groups of microorganisms leading to a "Qualified Presumption of Safety (QPS)" if the taxonomic group did not raise safety concerns or, if safety concerns existed, but could be defined and excluded (the qualification) the grouping could be granted QPS status. Thereafter, any strain of microorganism the identity of which could be unambiguously established and assigned to a QPS group would be freed from the need for further safety assessment other than satisfying any qualifications specified. Microorganisms not considered suitable for QPS would remain subject to a full safety assessment.

Application of genetically-modified bacteriocin producer strains on foods may be affected by different EU regulations and Directives. Strains modified by naturally-existing procedures (such as DNA transformation or plasmid conjugation) could be applied in foods with no other restrstions than those specified in previous paragraph for naturally-occurring strains. However, application of strains modified by procedures involving extensive DNA manipulation and artificial transfer to recipient cells is under much more strict control, including the specifications and limitations established by Directive 2001/118/EC (European Parliament and Council 2001) on the deliberate release of GMMs into the environment, Directive 2009/41/EC (European Parliament and Council 2009) on the contained use of genetically modified microorganisms, and Regulation 1829/2003/EC (European Parliament and Council 2003) concerning the marketing of GMOs intended for food or feed and of food or feed products containing, consisting of, or produced from GMOs. Bacterial products such as fermented bio-active ingredients prepared from GMM should also need approval in accordance with specifications under Regulations1829/2003/EC and 1333/2008/EC (European Parliament and Council 2003, 2008a, b).

References

Adams M (2003) Nisin in multifactorial food preservation. In: Roller S (ed) Natural antimicrobials for the minimal processing of foods. CRC Press LLC, Boca Raton, FL, pp 11–33

Aymerich T, Picouet PA, Monfort JM (2008) Decontamination technologies for meat products. Meat Sci 78:114–129

EFSA (2007) Introduction of a Qualified Presumption of Safety (QPS) approach for assessment of selected microorganisms referred to EFSA. EFSA J 587:1–16

EFSA (2008) The maintenance of the list of QPS microorganisms intentionally added to foods or feeds. Scientific opinion of the panel on biological hazards. EFSA J 923:1–48

European Commission (1988) Council Directive of 21 December 1988 on the approximation of the laws of the Member States concerning food additives authorized for use in foodstuffs intended for human consumption (89/107/EEC). Official J L40:27–37

European Economic Community (1983) European Economic Community Commission Directive 83/463/EEC. Official J 255:1–6

European Parliament and Council (1995) Directive No 95/2/EC of 20 February 1995 on food additives other than colours and sweeteners. Official J L61:1–53

European Parliament and Council (1997) Regulation (EC) No 258/97 of 27 January 1997 concerning novel foods and novel food ingredients. Official J L043:1–6

European Parliament and Council (2000) Directive 2000/13/EC of 20 March 2000 on the approximation of the laws of the Member States relating to the labelling, presentation and advertising of foodstuffs. Official J L109:29–56

European Parliament and Council (2001) Directive 2001/18/EC of 12 March 2001 on the deliberate release into the environment of genetically modified organisms and repealing Council Directive 90/220/EEC. Official J L106:1–38

European Parliament and Council (2002a) Regulation (EC) No 178/2002 of 28 January 2002 laying down the general principles and requirements of food law, establishing the European Food Safety Authority and laying down procedures in matters of food safety. Official J L31:1–37

European Parliament and Council (2002b) Commission Directive 2002/72/EC of 6 August 2002 relating to plastic materials and articles intended to come into contact with foodstuffs. Official J L220:18–58

European Parliament and Council (2003) Regulation (EC) No 1829/2003 of 22 September 2003 on genetically modified food and feed. Official J L268:1–22

European Parliament and Council (2004) Regulation (EC) No 1935/2004 of 27 October 2004 on materials and articles intended to come into contact with food and repealing Directives 80/590/EEC and 89/109/EEC. Official J L338:4–17

European Parliament and Council (2008a) Regulation (EC) No 1331/2008 of 16 December 2008 establishing a common authorisation procedure for food additives, food enzymes and food flavourings. Official J L354:1–6

European Parliament and Council (2008b) Regulation (EC) No 1333/2008 of 16 December 2008 on food additives. Official J L354:16–33

European Parliament and Council (2009) Directive 2009/41/EC of 6 May 2009 on the contained use of genetically modified micro-organisms. Official J L125:75–97

FSIS, Food Safety and Inspection Service (2002) Safe and suitable ingredients used in the production of meat and poultry products. Directive 7120.1, Washington, DC

US Food and Drug Administration (1999) Federal Food, Drug, and Cosmetic Act, Washington, DC

Wessels S, Axelsson L, Bech Hansen E et al (2004) The lactic acid bacteria, the food chain, and their regulation. Trends Food Sci Technol 15:498–505